Call Me Red

Call Me Red

A Shepherd's Life

Hannah Jackson

with Will Millard

EBURY
PRESS

1 3 5 7 9 10 8 6 4 2

Ebury Press, an imprint of Ebury Publishing
20 Vauxhall Bridge Road
London SW1V 2SA

Ebury Press is part of the Penguin Random House group of companies
whose addresses can be found at global.penguinrandomhouse.com

First published by Ebury Press in 2021

www.penguin.co.uk

A CIP catalogue record for this book is available from the British Library

ISBN 9781529109115

Printed and bound in Great Britain by Clays Ltd, Elcograf S.p.A.

The authorized representative in the EEA is Penguin Random House Ireland,
Morrison Chambers, 32 Nassau Street, Dublin D02 YH68

To Nan and Fraser.

For the woman who guided my dream
and the dog that got me there.

Contents

Prologue
Call Me Red

'She designed a life she loved' – **Anon.**

'Red. Take the centre. Up on the fell top.' Shoddy is delivering me his instructions with relative calm at the moment, but we all know his rage burns real bright just below the surface. He will be tearing his lungs into me within a couple of hours. It's near guaranteed.

I don't even know how he can see me from all the way up there in the clouds. I must only be a tiny little dot, a red-tipped matchstick hidden among the great stack of grasses and rock, but Shoddy has a powerful instinct for cock-ups, real or imagined, and possesses a poet laureate's creativity when it comes to his enthusiastic overuse of the f-bomb. He is the king of the sheep gather, the person with more experience than all of us put together, the elder statesman who we will all be deferring to when any major decision needs to be made, but everyone up here is highly likely to experience the sharp side of his tongue at some point today. You can hardly accuse Shoddy of being prejudiced: he hates us all equally. You'd be forgiven for thinking he emerged swearing from the womb with a flat cap on his head and a sheepdog by his side.

It's the summer's end in 2020; time to gather all the sheep off the high fells and bring them down to the lush grazing pastures in the lower valleys. There are five farms up here today, meaning there's five farms' worth of sheep spread out across miles of these hills, five farms' worth of marks to decipher on the sheep's backs (when we do eventually get them all down) and, most challenging of all, five farms' worth of opinion on how the gather should be executed.

There is always a bit of tension in the mornings before a big fell gather. The memory of all the stuff that's gone wrong here in the past. The hurdles, the scars and the big black holes that can suck sheep into their clutches for hours. All the physical snags that Shoddy knows instinctively from a lifetime spent in this area, and expects us to know too, as if we had already had his shepherd's map of the land hammered into us by our fathers, and their fathers before that, and their fathers before that. Instead, the reality is that I've just rocked up here fresh for a day's work as a contract shepherd: a sheep-farming mercenary out for another paid gig somewhere on the high fell.

You get used to this on contract jobs. It can feel a little intimidating at first, but what the lads from all these local farms might have over me in knowledge of this particular patch, I make up for in pure grit and determination. Not only that, but I gather across multiple fells throughout the year, right across Cumbria and into the Lakes, from the end of June right through until November, so my experience is on par with theirs. They may well have seen every challenge this piece of fell can throw up, but, believe me, my dog and I have also seen it all, just in different places.

I'm only 28, but I already have my own farm and my own animals to care for when I'm not out on jobs like this. Sheep, pigs, goats, chickens, ducks and a horse; all waiting for me down on my own fields in Croglin. Even if you were from an ordinary farming background it would have taken a steep learning curve to get into my position at such a young age, but I grew up in a working-class terrace in the middle of urban Wirral and didn't even set foot on a commercial farm until I was 20. You would have struggled to chuck a ball across the yard in my first home, let alone keep sheep, and yet here I am now: confident and secure with my job, my relationship and my dog, high up in these uncompromising hills. It's been one hell of a journey, though.

It's still the last week of summer, but today it may as well be November. It's going to be wet. The kind of wet that soaks you right through and should ordinarily cancel these big gatherings, but this one has already been postponed twice and everyone is getting itchy paws and wellies. Impatience isn't a reason to get worked up, though. Really, we should have all stood down for one more day. The forecast is glorious tomorrow, but this morning, I couldn't even see the fell out the back of my own farm.

The hills in this part of Cumbria pull in rain clouds and nail them down all day. No matter how hard the wind blows that rain will not be shifted from its tether, but it's not about how cold or wet you get, it's the lack of visibility that makes these days an absolute grind. Filthy low cloud bases are capable of hiding entire flocks of sheep from me and my dog. This is the sort of day that separates the men from the boys, the dogs from the shepherds, and the sheep from every single one of us. If you can avoid gathering on days like this, then you definitely should, but Shoddy can't do tomorrow. He said at the meet-up that we should just gather tomorrow without him. 'I won't be around forever,' he joked. But we all know what the punchline of that one is: by all means crack on without me tomorrow if you want, but if you make a mistake, and I'm not there, then it is all on every single one of your heads. So, I'm annoyed about that, I'm annoyed about the weather and, to cap it all off, my lift has just cried off, so I'm having to walk the entire fell before I've even started work. Shoddy isn't the only one who can dish it out. If I get crossed today then they're all getting it. I pull my jacket up tight round my neck and stomp off into the mist.

Here, they just call me Red. It has been a battle to get to this point, but when I cross that line and join the gather I'm no longer a townie, a Scouser, or a woman. I'm the Red Shepherdess: the hard-worn, bad-ass contract shepherd and farmer with Fraser, the top dog that everyone respects, at my heel. No one in these hills today would dare ask me: 'So, what's it like to be a female shepherd?' And yet that is the very question I have had to answer time and again through every single step of my career so far.

We push into our section and spot the sheep scattered randomly across the entire hillside. It's like someone has thrown a handful of rice up into the wind and left it all where it fell. They are spread out everywhere. 'Come by, Fraser.' I command him firmly without needing to shout. He is desperate to go at them.

I don't know what I'm going to do without him, but this year I know I have to face one of my hardest decisions ever. Fraser has been showing signs of slowing down. No one else round here could tell on sight, especially on days like this, but the depth of our relationship means I can feel it, and so can he.

Leaping forward, Fraser sprints up the left side of the hill, outflanking all the sheep at some serious pace in a single move. He keeps just enough distance not to freak them out, while circling them in a counter-clockwise direction, squeezing them gently together into a group, like an expert fisherman corralling a giant shoal of sardines into a purse-seine net that the fish didn't even realise was there.

This is called the outrun. You could stick a shepherd here from four hundred years ago and they would know exactly what was going on. I wonder if they might be curious about why the person responsible for this section of the fell is a lady? Four hundred years ago, I'd hope they had bigger things to worry about.

Do you think these hills or animals ever cared that I'm a woman? When it's minus-10 in winter and a sheep is stuck in deep snow, and Fraser and I have to go and save its life, does it check my gender first? Or when a lamb is locked in a breech-birth position and about to kill its mum, is that when the extra X in my chromosomes matters? Believe me, nothing screams equality quite like bending to your knees, alongside the rest of the farmers, to clear maggots from the rear-end of a wounded ewe in the height of summer.

'Stand!' This is a critically important moment. I need Fraser to maintain his distance without letting the sheep scatter. I know he will – we've done this thousands of times before without fail – but the wolf-instinct is strong in the best of the sheepdogs. Being able to gently lift the lid on that primal box when I need it, and then firmly slam it shut when I don't, is the difference between having a good day and an absolute nightmare. He is about half a mile from me, but I can feel his slightly overeager posture. He impulsively wants to charge at the sheep now he's brought them all together in a neat group, but if he does that they'll spread all over the hillside again and Shoddy will go absolutely nuclear.

'Wait. Steady. Walk on,' I repeat, teasing him in and out in stages, like a black and white yoyo I'm holding on an invisible string. Gently, I allow Fraser to drive the sheep along in the right direction. He's an expert at this and at settling into the rhythm I'm setting, right to the point my calls just become a background reassurance.

'*Man's* best friend.' What a joke. We have the closest bond I could ever imagine between a human and an animal. It's symbiotic and more than just a partnership between workers. He's not my 'right-hand man', he *is* my right hand, executing my exact thoughts the moment I have them, moving our sheep along unseen lines in the land that exist only in our shared mind's-eye.

Today was the first time I ever questioned whether I should let him lead the gather. Really, it's high time I pushed his protégée, Butch, into the top-dog position and relegated Fraser to being a highly experienced, but second, dog. I've brought Butch up here for the first time and she (yes, Butch is a she – don't ask, you'll find out later) is visibly shaking with her deep yearning to get to work. My partner Danny is up here for the first time too. He is buzzing with the thrill of it all. He can't stop smiling in spite of all the crappy weather, and keeps trying to make me laugh and lose my focus. I scold him gently, but I love it all really.

Squall after squall comes rolling in from the direction of the Lake District but we are breaking the back of this job now. I know, in my heart, that Fraser's retirement is inevitable. No dog can last forever in this hard and uncompromising environment, no matter how elite their DNA is, or was. But I just can't face staring into his deep dark eyes, framed beneath those beautiful cocoa patches of fur, and telling him really that, after everything we have been through, 'This is it.' He's as much a part of me now as my red hair.

We go after more sheep and I restart the calls from the very top of the ridgeline this time round. Danny stands by my side, face-first into this wild weather. Sometimes I can't believe I chose to work in this particular office, and yet I have never felt so at home. It's like a part of this place has always been carved into my heart. There is nowhere else I'd rather be.

Prologue

I first dyed my hair red when I was 17 and got my nose pierced on the same day. My poor dad took me to get it all done. I couldn't articulate it back then but I somehow knew that it was more than your average act of teenage rebellion. I think Dad knew it then, too. That I was dyeing my hair and choosing to be free.

When you're 17 and a woman growing up on the Wirral it feels like your path in life has already been laid out for you by someone else. This red hair represents a rejection of all those things I was supposed to do. The rigid list of jobs I could work in. The rigid sets of friends and boyfriends I could hang out with. The rigid sets of clothes, looks, houses, cars and very occasional holidays I could choose from. I knew then that it was all bullshit, but I could never have known that this windswept place was where I would truly find myself.

Today this red hair is *still* a rejection of all that mainstream society expects from young women who come from towns populated by old ideas, but it is also about continuing to choose to go against the flood of people emptying from the countryside too. I wish everyone knew that out here you can really set your own goals and standards, chase your own dreams, be your own boss, and judge your success not by the money in your bank account or the car you drive (although I would really like a new off-roader, please) but actually by how much happiness you can achieve for yourself.

Dyeing my hair red was, and is, the expression of the purest kind of freedom I could imagine. 'R. E. D.' I know it sounds cheesy but to me it stands for being 'Real Every Day'. They can put that on my gravestone: 'Hannah Jackson. The Red Shepherdess. She was Real Every Day'.

The attitude expressed in this hair might've pushed me

forward on this journey but I couldn't have done it without Fraser. He's somewhere down there in the mist, acting on messages and instructions hard-learned through the million small improvements, and epic failures, we've endured in our eight years together.

In spring he sat down in the middle of a job. I couldn't believe it. Right as we were moving sheep on a road from one field to another. He just gave up and let them all go past. I had to chase them up the road and turn the entire flock around by myself. He would never, ever, have done that before. My farm is getting bigger this year though, and so are the challenges. I know I need to start investing my time into training the other dogs and getting them up to speed. I suppose I need to keep training up that big daft bugger Danny as well.

All our sheep are heading out in a neat and orderly line across the fell now, with Fraser right behind, jostling them along. You can't see it, but just the other side of that hillock are all the sheep pens, waiting to gather them in for the big five-farm sort. Shoddy and all the other farmers are closing in on that point now too. Another job well done and perfectly executed. Not that we will be getting any praise from Shoddy, of course.

Long red hair whips in and around my eyes as I make my way back down the slopes, tucking in behind the flock too. I can feel emotion swelling up in my chest. If I'm going to find the strength to retire this dog then now is the time I need to remember all the reasons my hair *is* red, and everything that I have gone through to get us all to this point.

Today, they might call me Red, but this is the story of how I got here.

Chapter One
Dr Dolittle

I was born on 31 July 1992. It's a fair bet to say it was raining (it always rains on my birthday) but obviously I can't remember, and neither can my mum, as I'd effectively just broken her back in half. That special day, on the arrival of her first-born, my mum wasn't able to pick out the weather outside Arrowe Park Hospital because she was on enough opioid painkillers to transport her to the moon.

I hadn't been planned. Mum and Dad – Mandy and Stuart – had barely been married a year when she fell pregnant at 20 years old. Money was tight. Dad was earning very little, while trying to pull his father's photography business out of debt, and Mum was only earning slightly more working in the research labs at Unilever. They probably could have done without an extra mouth to feed, but Mum always says: 'Hannah was a nice surprise.' That tells you everything about the depths of kindness I've always received from my parents. In reality my birth would have added huge financial pressure, and that's without considering the nine months of extreme heartburn Mum endured during her pregnancy with me, but they never ever complained. Ultimately, I rewarded her endeavours by pushing two of her spinal discs clean out of her spinal cord. She just says it was an early sign of how determined I was to enter the world. Like I said, my parents are incredibly kind.

My mum comes from a long line of very tough people. Every time I've had to endure the discomforts and pain that come with my profession I know my strength has been part-gifted to me from my mum and everyone else that came before her. She does actually have farming on her side. Her grandad Bill, my great-grandad, lived on a family farm over in Ireland. Times were exceptionally tough in the countryside back then, and Bill would drive his animals on foot for many hours to make it to the local markets. Often they wouldn't sell, which meant he had to return the entire distance home with his livestock, empty pockets, and an even emptier stomach.

The stress and uncertainty eventually became too much for Bill's father, who tragically took his own life, but great-grandad Bill was a man of quiet pride and deep caring. It wasn't actually until the final stages of his battle with cancer that he ever even spoke of his father's death to the family, or the fact that he had discovered his body, or even the weight of responsibility he must've inherited as the new man of the house. His sudden urgent need to provide for his seven siblings saw him sent alone to England, where he ultimately found work as a labourer on many hundreds of different building sites, during a long career that lasted into his seventies. It marked a pause in my family's agricultural endeavours, until I took up the baton many decades later; but I learned through his story that, when my family have achieved, it has often been with their backs against the wall and a stoic determination to build a better life.

There is often a strange symmetry to life when you truly start looking for it. In researching for this book I discovered my boyfriend Danny's father had actually worked with Bill. He couldn't have known of the extent of Bill's struggles in his youth, but he did call him 'The Concrete King', and mentioned how proud Bill had been of his six children and all their achievements. It transpired he had actually chosen to work under Bill: 'I was always with the best, because I think if you listen to the best, you learn.'

Bill set the bar unbelievably high with his work ethic, grit and basic bloody-minded survival instinct. As much as I hope no one in my family ever has to endure what he went through, I can't help but find his story deeply inspiring whenever I think about him.

Having thoroughly beaten up my mum, I eventually came into the world with a head so badly disfigured that I must've looked like I'd been pulled out with an industrial sink plunger. It resembled a swollen shoebox; all rounded and bloated, with these weird-looking sharp corners sticking out at the back of my skull.

Basically, if my mum had been a sheep, after producing an animal that looked like me (toothless and hairless until I was one year old, by the way), she definitely would have been culled. It turned out it was just me, though. My two younger sisters, Holly and Caitlin, both arrived in later years with relatively little upset and, as far as I know, normal-shaped heads.

I began life in a tiny terraced house in New Ferry, an urban area that hugs the muddy banks of the River Mersey, right over on the eastern edge of the Wirral Peninsula. The house sat just off the Rock Ferry bypass, which funnelled the traffic up the Wirral and down into the road tunnels that feed the grand city of Liverpool.

Our street was blocked off at one end, so the traffic was never too bad, but we lived in close quarters with each other and all of our neighbours. There was very little jealousy or 'keeping up with the Joneses' going on though, as everyone on our street had roughly the same: not a lot.

Our place was fairly typical. A pebble-dashed façade, with a sitting room at the front, a kitchen-diner at the back, two bedrooms and a box room upstairs, and a really steep set of open stairs. It was all single-glazed, so it got really cold in the winter months, and we only had a small concrete-covered yard out the back, so the possibilities for exploration were a bit limited. You probably couldn't get further in your imagination from the wide-open and rolling Cumbrian countryside where I live and work today, but it was our home, and our home was always filled with happiness.

My parents had even less money now Mum was off work and looking after me, plus interest rates on mortgage repayments were sky-high at the time, so even little places like ours felt excruciatingly expensive. I don't recall ever going without, though. I didn't know that my toys and clothes were mostly second-hand or from charity shops, or the hugely supportive behind-the-scenes role that Mum's mum, my amazing nan, Margaret, was having. She helped out with childcare and frequently paid for all those 'little extras' that can completely cripple low-income families. Ultimately, the biggest reason I didn't feel like I had missed out was because what children really need isn't actually found on shop shelves anyway. Love, time and security were in super-abundance in our house. My parents always put their children first.

By today's standards the area would be considered 'deprived', which immediately throws up some illusory Dickensian scene with poorhouses, open sewers and never-ending antisocial behaviour. Admittedly, it did take my mum, with her innocent Catholic upbringing, about 18 months before she realised the red light shining from the top window of the house opposite wasn't because our neighbour was keeping a menagerie of nocturnal reptiles, but truthfully our place was a lovely area to grow up. The houses may have been small but the people that lived in them owned them for themselves and cared about them deeply, just as they cared about their neighbours. You could say what you liked about the area, but the people of New Ferry genuinely looked out for each other.

There is this myth that people from towns and cities don't know about real community. That myth is especially powerful in the countryside, where some rural people like to think they have a monopoly over what a community should look like, but it isn't true or fair to say that people in urban areas have no concern for each other or their surroundings. In the so-called 'have not' parts of the Wirral you can actually find the strongest and tightest bonds of anywhere you could care to look, and it is that pure sense of belonging that keeps people living in places like ours. Despite the aesthetics of the houses and streets, and the relative lack of work opportunities, people hold on there not because they are poor, lazy or just lacking in ambition, but because they actually really feel part of something.

The truth is there is good and bad everywhere, and you shouldn't judge anyone just as a product of where you think they are from, or what you think about those sorts of places, especially if you've never been there for yourself. One of the most important lessons my parents drummed into me from a very early age was that everyone should be treated equally and taken at face value. Always.

As long as I can remember I have felt restless if I'm not actively doing something, and especially if I'm cooped up indoors. I started crawling early, and by nine months I was able to climb right up and out of my cot. By three, I could do the monkey bars. Apparently I didn't like sleeping that much, which must've been an absolute nightmare for my parents, but the ability to find a way to function on just a few hours' kip has helped me enormously as a farmer today. Sometimes, and some seasons, you've just got to be able to draw down on your reserves of energy, even if you've had less than half a night's rest.

My parents both worked incredibly hard, pretty much every hour God sent them, but time for the family was always worked in somehow, probably by sacrificing time alone for themselves. In fact, they've only recently had a proper holiday together with just the two of them, and that was their twenty-fifth wedding anniversary. Their work commitments – by now Mum was helping Dad in the photographic business – often meant we would all be packed up together: me, my sisters, and whatever animals we had in the home (that I absolutely insisted we couldn't leave behind), and taken with them up and down the country. It meant as soon as my parents were done with satisfying the demands of whatever client they were meeting, they were then able to switch their sole focus onto us, the second they closed the office door. In that way I was fortunate to see and experience so much more of the country than I otherwise would have done, and I probably grew much more comfortable with the idea of being a little bit nomadic, something that would also help me considerably in later life.

One of my earliest and most treasured memories was the morning Mum came home with a VHS copy of *The Lion King*. This was extremely exciting as, firstly, it was brand new, it had literally just hit the shelves that morning, so that felt like an extraordinarily spontaneous and almost reckless purchase; and secondly, because my dad, who had never ever missed an hour of the working day, downed his photographic tools to curl up with me and watch it in its entirety, from its start on Pride Rock with the birth of Simba, to its finish, on, er, Pride Rock, with the birth of Simba and Nala's cub.

I've probably watched it several hundred times since, upgrading that film from VHS to DVD, to music CD, and on to whatever streaming platform I'm using today, but the feeling of immeasurable calm and sense of belonging I felt in that moment, cuddled up to my dad and immersed in this wonderful world of African animals, was absolute. It bubbles up lovingly from deep behind my belly button every time that sun rises over the African plains.

I was captivated. The messages about family, the relationship with parents, friends and loyalty, as well as this extraordinary cast of cartoon – and fully communicating – animals, had a profound influence on me. My fascination with the wild outdoors had always been there, but this film articulated at a very early age the way I was already beginning to see the world. I might have lived in a little terrace on the Wirral, but I fully believed in the idea of this huge magical theatre filled with the most wondrous cast of creatures, that were just waiting to meet me.

As my first words really started to flow my family started calling me Dr Dolittle. It wasn't quite enough for me to just passively observe nature from afar, I wanted to get close, touch and, ideally, actually hold it and communicate with it.

There was never a dog that walked down our street that I didn't feel an instant magnetic attraction to, but it didn't stop there. I would touch any animal, or even anything that resembled an animal. I was the kid that stuck her hand higher-than-high to get to feed the goats at the urban farm, I would leap to volunteer to stroke a rat, or handle a snake during our frequent visits to Chester Zoo. I would pick up anything – snails, worms, shells, leaves – and examine them with forensic interest. Even today, I'll rarely be conducting a conversation without rolling my fingers across a twig, or separating some strands of fleece, or, more likely, pulling up a shoot of grass and, ever so carefully, picking out the individual blades held within the tiny sheaf that binds all the grass leaves together, until I'm eventually just left with the tiny bulb of a grass root.

I could never contemplate deliberately or unnecessarily harming an animal, though. I wasn't like those horrible children that take pleasure in pulling the legs off flies. I just had this deep interest in the component parts that formed this thing of absolute beauty in front of me. I guess I inherited it from Dad, who was forever figuring out how a piece of technology worked by taking it all apart and putting it back together again. As a farmer, the ultimate evolution of that interest has given me an eye for the very smallest of details. An ability to actually see the wood for the trees, to instinctively isolate struggling animals from otherwise entirely healthy flocks: a lamb with a broken leg, a single chicken with missing tail feathers, or just a single wary look from a single dog that might indicate they had a problem. Last week, Fraser cut his face on a gather for the very first time. It was just a nick, but I could sense it on him from 30 metres, long before he was close enough for me to see the claret blood running through his eye patches. In my job, it's the tiny details that matter, and you ignore them at your peril.

Unfortunately, as a child, my attention for detail and total fearlessness led to me meeting one microscopic creature that I really wish I had never encountered. Campylobacter, the spiral-shaped bacteria that inhabits the intestines of many warm-blooded animals, most probably made its way into my body when I had accidentally, or otherwise, touched and then ingested dog poo.

I was only aged four when I was taken into hospital for a week. It wasn't my first trip inside, and it certainly wouldn't be my last, but in terms of the severity of my sickness it was almost certainly my worst. My parents were beside themselves with worry, particularly as it took several days before the doctors and consultants had even figured out what was wrong with this slight and slightly feral-looking child, who just could not stop vomiting.

I became dangerously dehydrated, my weight and fluid loss eventually becoming so bad I was informally prescribed fizzy drinks, just to get some sugars and water back into my body. The fact I also ripped out and hid the wide-bore cannula connected via my arm to the emergency drip certainly did not aid my increasingly poor state, but eventually, with the correct diagnosis, careful monitoring, and the occasional broadcast of *The Little Mermaid* on the ancient TV that was periodically wheeled into my ward, I bounced right back.

If this was supposed to be a hard lesson in modifying my childhood behaviour, it was to have absolutely zero impact. If anything, I got worse. The next year I was back inside again, this time with a fractured skull from roughhousing in the playground. Initial assessments focused on the extraordinarily strange corner-shaped lumps on the back of my head, until my mum put them off the scent with the description of my skull from birth. The fracture they eventually found, despite receiving the blow on one of my beloved skull corners, was actually right across the front of my head.

I grew up and was soon refusing to wear clothes; despite

Mum's best efforts, I remained virtually shirtless until the age of ten. It wasn't down to a lack of parental care that these sorts of things were happening, my mum could control all of us with just a look, it was just that I was resolute in wanting to live my life my way. As much as my parents would do everything they could to keep me from harm, they also knew the importance of giving their children space and time to develop their own identities and ideas; even if there was an inevitable degree of risk in me doing so.

In keeping with my developing 'tomboy' attitude, I had my hair cut short and gravitated towards playing football and skateboarding, building dens, and getting dirty hands and scabby knees. I had no idea that what I was doing wasn't exactly stereotypical behaviour for girls my age. Even if I had, I probably would've done it anyway, but now I appreciate how lucky I was to have parents who allowed me the freedom to express myself completely from a very young age. I never once felt pressured by the societal ideas of what young girls were 'supposed' to be into (long hair, pink princesses, doll's houses, etc.), in fact, thanks to Mum and Dad, I wasn't actually aware such obligations existed at all. My two sisters were much more conventional, but that was through their choice too. Being able to mix it up in a so-called 'man's world' definitely helped when I began my journey into farming much later in life. I wasn't ever doing it to prove that women could, I was doing it for the same reasons I picked up all the wildlife, did the monkey bars and kicked a ball: it was simply what instinctively made me happy.

It's always other people that make such a big deal of what girls and boys, men and women, are supposed to do, or not. Take that away and actually just let all people follow their own passions and I have absolutely no doubt you'll soon discover a healthier and happier society for all.

I find it really sad when you see boys and girls of all ages stuck indoors anyway. There should be no barriers to children from any background building on an interest in animals. Granted, it is a lot more challenging when you grow up in the middle of a town or city, but it still isn't impossible, just look at me. I feel that people, and parents especially, get hung up on the idea that the countryside and the outdoors is too dangerous and dirty for their children to just discover it for themselves. I see it all the time, this obsession with cleanliness and sterility that suffocates any chance of self-discovery and natural childish development, pretty much from birth.

Okay, I know you could point to the fractured skull and campylobacter incidents as not the most brilliant examples to support my argument, but not only did I survive my childhood, I turned those things I loved the most into the job I absolutely adore today. My mum frequently reminds us what my health visitor once said to her: 'There's nothing wrong with your kids learning to write in the dust.' She's right, and I guarantee you there are many more miserable children supposedly safely indoors but actually stuck in front of really harmful video games and screens than there are those out there learning about the world by getting on their hands and knees and picking it up with both hands.

Just avoid the dog poo and your kids will be fine.

Mum and Dad eventually managed to consolidate all the debts of the photography business and started a new arm of the company. It aimed to take a punt at investing in this all-new digital image technology that was starting to emerge in Britain around the mid-nineties. It was a masterstroke of forward-thinking on their part; the digital era would soon take over and, ultimately, it changed our fortunes as a family completely.

That business became so successful it was eventually bought out, and suddenly we were in a position to grow and build. We even had enough financial security to move house and welcome my second sister, Caitlin, into the fold.

The new house in Bebington was still around the same size as the one in New Ferry and was actually only a mile away to the west, but the change was massive. It was semi-detached with room to extend across a big garden, and even had a drive. Better still, it backed on to a massive playing field and cricket pitch, which stretched right across to my new primary school. Quickly, our house became the de facto headquarters for all my new friends and me to play our endless outdoor games, always rounded off with chip butties bought by Mum from the local fish-and-chip shop.

More space meant more pets. There was no nature table, ours was a nature house. Alongside all the dogs (I have never not had a dog at any point in my life) there is a roll call of childhood pets so long it is nearly impossible to recall every critter that graced either house in New Ferry or Bebington. Some were acquired conventionally, but many others were rescued (whether they, my parents or, occasionally, their rightful owners liked it or not).

At the entry level there were sea-monkeys, a type of brine shrimp and popular novelty pet that was never likely to hold my attention for too long; these were followed by Ant and Dec the goldfish, tropical fish of various species (but mostly anything that vaguely resembled a shark or had live babies), injured hedgehogs scooped off the road, and tadpoles and froglets 'borrowed' from ponds. I should mention here my first dog, Tilly, the cavalier King Charles spaniel. Poor Tilly, she accidentally discovered my mum's back tablets and died, so I feel partly responsible for that one. Then I graduated to Lemon and Lime, my parrotlets, a mini-parrot of South American origins. I can't not mention Sneezy the rabbit, he was the first in a long line of rabbits (unfortunately for Mum and Dad, our first house was directly behind a pet shop, shout out to Andy's Aquatics, who also sold rabbits, obviously). Sneezy had a special place in my memory, as, one morning, I came down the stairs and he had been swapped by what Mum and Dad had naively presumed to be an exact replica that I wouldn't notice. I don't know exactly what had happened to cause Sneezy's demise, but even at four I could instantly tell that this impostor was definitely not him . . . but I kept him anyway. We also had Chloe the cat and, later, Meg the golden retriever, who used to get me off primary school lessons by running around the playground until my teachers told me to take her home.

Rosie was our first collie, then there was Hamish the ginger cat, born feral in a car park near Mum and Dad's work. After a holiday in Cornwall, we came back with two more adopted rabbits: Rubens Barrichello and Schumey (we are big Formula One fans). Sadly we lost Schumey to a fox, but we soon had Rubens Barrichello so tame he actually used to live in our Wendy house, and was frequently discovered in the wooden play microwave by Mum, happily passing the time. There was another adoptee, 'Little Mo' the cat (who actually abandoned me, due, I think, to jealousy, when I rescued another cat, Margo, from a pub), then another rabbit, Mickey Blue-eyes, then Todd the near-uncontrollable Border collie, then Dan, Todd's half-brother.

Before that there was 'Pure-Evil' the hamster, who I trained until he was just 'Pure', before he was tragically overfed by my nan to the point he actually got stuck in his play tube and died. Oreo the African pygmy hedgehog did the smelliest poos of any animal I have ever kept, ever, and finally, probably the most radical rescue was the pair of lambs intended for slaughter, Greta and Gunnar, which I got my mum for Mother's Day. That pair brought legions of kids into our Bebington house-yard-cum-amateur-farmyard-zoo for many years to come.

Incredibly, Mum is actually allergic to animal hair. Looking back, her tolerance for my obsession was absolutely extraordinary. She always knew the importance of animals to me, and my parents knew the importance of allowing me to learn the life lessons to be had from the responsibility of looking after a living creature. I just don't think they could possibly have known just how many animals that germ of encouragement would bring into their home, or even where it would eventually lead.

I sometimes wonder what it might be like if, many years after the memory of our family and all the pets we had has faded, someone decided to re-landscape and dig up our back garden in Bebington. I imagine it'll be quite a shock when they discover dozens of animals, across all species, ritually buried in everything from shoe-boxes to Smarties tubes – but those animals were (mostly) as loved as we were, and every single one was special.

I'm not daft enough to think I can actually have direct and complete conversations with animals in the same way as the fictional characters in *Dr Dolittle* or the highly Disney-fied *Lion King*, but I don't believe that animals are just inferior or stupid either.

Having so many animal encounters in my childhood established early on that conventional wisdom was not always correct. I spent a great deal of time trying to understand the animals in my care – I still do – and it was always a surprise to other people how infrequently our pets would fall out and actually fight each other, especially given how many were supposed to be sworn enemies. Honestly, animals do not act on knee-jerk impulse or an uncontrollable instinct all of the time. The cats and dogs generally got along, the animals saved from their rough beginnings would nearly always flourish, and even the smallest of our creatures, like the hamsters, were left alone or even played with by the other larger animals, on many occasions.

This isn't to paint some over-romantic idea of all our animals and us always living in a state of perpetual harmony – again, real life isn't Disney – but it is fair to say that there is often a far greater depth to animals than we credit them for. If you can find the time to give an animal your focus, care and attention, you might be surprised at the level of understanding you can find. In some cases, you can form a deep bond that is so tight you are almost able to communicate your will and even collaborate. But you do need to believe that there is something really special, and worth discovering, deep down there in the first place.

Chapter Two
Margaret

My nan filled the room with love. Even if she'd only seen you the day before she would hug you as if you had been apart for a hundred years. She was tiny, but she had the most enormous hair. It was this luxurious voluminous bouffant with tight ringlet curls and it was always going to be the first thing you noticed whenever you saw my nan.

Her skin was given much attention to detail too. She was young to be a grandmother anyway, but Nan never wanted to appear old. The Shopping Channel was always on and a procession of various skin creams, all guaranteeing the elixir of endless youth, filled her dressing table. She wore so much anti-aging cream her face shone so brightly you could've spotted her in a blackout, and she also used this funny little TENS machine with tiny stick-on electrode pads, which gently shocked her facial muscles, guaranteeing they were always in tip-top condition.

She always wore these super-shiny, floaty-looking, eighties-style clothes too. They often had a Day-Glo fluorescence and the sort of colourful flower patterns you'd see on old people's curtains or couches. I thought they looked absolutely horrendous, but she adored them, and would twin her look with these flip-flop shoes that were usually gold, and always with a little heel. Even if she was in the house all day, Nan looked like she was waiting for the call to the 1983 Oscars. Even her slippers were heeled.

Behind it all my nan was naturally a very beautiful woman anyway. Her smile alone could turn heads, but she also physically radiated a type of beauty that only comes from having the purest form of kindness etched deep within your soul. She was the nicest person I have ever known, and with that, and in spite of her diminutive size, her sense of presence was absolutely colossal.

Nan was the life and soul of her community. She was universally loved and used to be very outgoing. If there was one thing you could say that summed her up: whatever needed to happen, especially if it was helping someone in any kind of need, it *would* happen if my nan was involved.

She was an extraordinarily generous person. Every week after Mass, she would visit all the sick in her congregation, bringing them their communion, and she would often cook and bring food for the homeless too. She was constantly on the phone ringing people that were ill or feeling down. I would hate to think what her phone bill must have been, as no one was ever forgotten or left behind. Nan was the shoulder to cry on, and a loyal counsel for a vast network of people. I never ever heard anyone say a single bad word about her. She had huge reserves of faith in humankind, and loved her family – my mum, my dad and us girls – above all else.

Margaret Madden was the second of my great-grandad Bill and great-nan Betty's six children, and was only 17 when she had my mum, Mandy.

It was a massive shock in Bill's strong Catholic family. He had come from a family of seven, and already had six himself, so Bill and Betty were no stranger to kids, but 17 was undeniably young. There were, after all, just five and a half years between his youngest daughter and his first grandchild.

My mum's biological father didn't really do that much in her early upbringing and was completely absent by the time she had turned six. It transpired he had other children with someone else, a fact he had kept secret from my nan, so their relationship was almost inevitably doomed from day dot.

Bill just did what Bill had already learned to do in his life: he cracked on with shouldering the extra responsibility of being a second father, as well as a grandad, to my mum. Love came easily for Bill, he absolutely adored my mum. He called her 'Little Bill' and was steadfast in his belief that she would most definitely become prime minister one day.

Mum and Nan had to stay in the childhood home together with the rest of Nan's siblings. All of them sharing one room, and Mum and Nan sharing one bed. That was how life was. Hard, but not quite as hard as Bill had had it back on that family farm in Ireland when his dad had taken his own life. It's almost unbelievable to consider now, but there was just an acceptance of life's hardships back then. As long as things were getting better, even if it was just by a tiny bit, and everyone was more or less fed, then the family just rumbled on forward.

Nan wouldn't have another child or meet another man until Mum was 14 years old. There was no time for that because, as soon as she could, Nan went to work to help provide for my mum. She even packed her off to primary school a whole year early so she could have extra hours in a company that manufactured blouses.

Nan was undeniably bright and (unsurprisingly, given she was one of Bill's children) incredibly hardworking. She worked all day at the blouse factory and cleaned three nights a week on top, taking my mum along with her in the evenings to keep an eye on her and, later, to help out.

Promotions came easily and, 25 years of graft later, she was in charge of the entire quality control and merchandising department for the blouse company. Her job involved a lot of travelling to maintain the quality standards nationwide, and that was how she came to have the first of two tragic accidents that would completely change our family's lives.

My nan was passenger in a work's van. It had only been serviced the previous day when the brakes seized on the rear axle – they could have been anywhere when it happened but, unfortunately, they were at full speed on the M56. The van spun hard across the three-lane width of the motorway, dumping them back-to-front in the outside lane, and directly facing the oncoming traffic that was thundering towards Nan and the driver at breakneck speed.

It was 7.00am, right at the beginning of rush hour on the major artery between Chester and Manchester. Their ordeal wasn't going to get any easier; in fact, it was just beginning. From the frying pan into the fire; they managed to escape the car only to find themselves trapped on the central reservation with six full lanes of traffic hurtling all around them.

The police arrived quickly but, utterly bizarrely, instead of just stopping the traffic to safely remove Nan and the driver from the scene of the accident, the officer in charge instructed them to hold his hand and sprint across the entire carriageway of the M56 east-bound. The police then turned the van around, drove it to where they were standing on the hard shoulder, and pretty much told them to crack on as if nothing had happened.

That was impossible as the brakes were still completely seized up, so there they were left, awaiting rescue, with traffic thundering past, rocking the van and Nan's tiny frame.

The driver was okay, but the incident really affected my nan.

It marked the beginning of what we now all know was PTSD – a post-traumatic stress disorder – that gradually saw this larger-than-life character retreat within herself and ultimately the confines of her home.

PTSD, if unchallenged, seeps into every aspect of your life, gnawing away in the background in your every breathing hour. The origins of the trauma itself, the story of the trigger – in my nan's case, the road accident – soon becomes secondary to the lived experience of the symptoms of this awful mental-health problem. All too soon the simplest of tasks, like just going to the shops or getting ready for work, becomes a terrific and terrifying ordeal. We are light years ahead with our specialist knowledge of PTSD today compared to then. Nan began to experience serious and debilitating flashbacks and, unfortunately, when she then took time off work to recover mentally, the insurance company that insured her work did not want to pay out her full sick-pay wages. Instead of being free to focus on just getting better, Nan was put under increasing pressure to return to her job and start travelling again, the implication being that she was fabricating her PTSD. Her symptoms quickly worsened after that but, as she started a legal case to try and get her sick pay, the insurance company then employed private detectives to spy on her every move, in an attempt to prove she was making it all up.

Not only was she physically followed – to the shops, to the hairdressers, to the church – they even filmed her through the windows of her own house. Given paranoia can be a common symptom of PTSD, it was just about the worst thing my nan could have had to endure in the aftermath of the accident. Regaining her small everyday freedoms were the incremental steps Nan needed to take towards her recovery. They were stolen from her completely in the pursuit of money. Inevitably her self-confidence took a severe and serious nosedive after that, and soon those reclusive tendencies that began as side symptoms of her trauma took over and dominated every aspect of her life.

She retreated from all interactions outside her home that weren't with her immediate family, and soon she was even showing signs of being uncomfortable in her own back garden. The curtains were drawn, or the blinds of her home were manipulated at such an angle that they could only let in light, and not prying eyes.

It was very sad to look back on, but as a child, it just became the norm and, if anything, it meant the focus of Nan's attention was moved even more sharply onto her grandchildren. Before the accident, she and I already had a special relationship that began with my birth.

The day I was cut from my mum, my nan was also deep under the surgeon's knife.

You could scarcely tell behind the big hair and oversized clothes, but Nan had lost an awful lot of weight in the months leading up to Mum's labour. Really, she had been quite ill a lot of the time; not that it had stopped her from working, naturally.

Eventually things came to a head and she ended up getting admitted to hospital. All the signs from her blood tests indicated something very sinister was wrong, and the resulting abdominal scans found a huge dark mass in the pit of her stomach. My mum went into labour just as Nan was laid up having investigative surgery. Mum had a cut right across her front to free me, and Nan had one straight up.

The day after I was born, and for every day Mum was recovering from her caesarean section, Nan was wheeled into the ward in a wheelchair for us to all be together. One of my most treasured photographs is the picture of my nan cradling me from her hospital bed. I can see her now, the picture of poise and beauty, with all her curly hair tumbling around her pillow. One hand cradles my feet, the other, seemingly full of IV tubes and bandages, rests across my left shoulder. My tiny hand is gripping her middle finger tightly and she smiles broadly behind the most tired set of eyes I have ever seen her wear. A Post-it note, hand-written behind her head, has just doubled her hourly intake of water, with the number 3 hastily and thickly turned into a 6 with marker pen.

She had been told that all the signs were indicating that the cancerous growth within her was obviously significant. That she would most likely require chemotherapy and lose her iconic hair. That her naturally positive outlook would help her, as would the arrival of her first grandchild, but really she should prepare herself for some battle ahead.

In the operating theatre that day, absolutely nothing was found. No specialist, no consultant, no doctor or otherwise, has ever been able to explain what that dark mass ever was, or where it had gone. Nan, with the understatement of the century, called it her 'little miracle'.

Nan had six weeks off work in recovery after her operation, which meant she had precious time with Mum and me. She took me for walks in the pram, dressed me in pretty pink dresses that I would've hated, but she also took me to the local zoo for the very first time, so I'll forgive her.

I am in no doubt the ordeal she faced down on the day of my birth bonded us together tightly. I was also very lucky that I had a handful of years with her before the road accident, and before the trauma of that really started to sink its teeth in and take over.

It used to drive my mum nuts, given how when she was a child she would've had to have scientifically proven to Nan that she was on the brink of death to be allowed even a single day off school, but Nan would keep us home at the tiniest sign of the most marginal discomfort.

If we were ever actually ill, and round her house, my sisters and I would be given a little bell to ring, so we wouldn't have to leave our bed, or really even lift a finger. If we ever wanted anything – and by anything I mean absolutely *anything* – we just needed to ring the little bell. A piece of toast? She'd bring you five rounds. A drink? She'd bring a jug. A square of chocolate or sweets? You'd be getting a bowlful. She also had this ointment called her 'magic cream', that she would apply for any given illness or ailment. It was a kind of pinkish candyfloss colour and was unfailingly applied to every one of my cuts, bruises, insect bites and stings. I clearly remember her also putting it on the open wound where my cannula entered my arm during the whole campylobacter episode. It was basically a low-grade over-the-counter antiseptic, but in my mind it really was magic, especially when it was applied with the tenderness of Nan. I'm pretty sure if I had my leg torn clean off today I'd be feeling right as rain after an application of Nan's magic cream, as long as it came from her hand; that stuff was Jesus's eighth miracle. She just loved to look after us and would do anything to keep us at her home; not that we needed any persuasion, my sisters and I absolutely worshipped her.

I mentioned before how she had quietly supported my mum

and dad during the tough times in New Ferry, but it didn't just stop once the business had been sold and we had moved. Anything she could give to help improve our lives and home, in any way, big or small, she did, and yet she lived so humbly and quietly herself.

We would have given up everything we had in a heartbeat to have seen her surmount just some of her problems; my parents, especially my mum, tried absolutely everything they could think of to try and turn things around. But the obvious fact was, the more she stepped back socially, the more she sought her life's pleasures by living vicariously through us all.

Nan was absolutely mad about Christmas and birthdays and started traditions that we will forever keep in our home. I can't begin to describe how many presents she would get for us at Christmas. Effectively we had a sack each, but all the gifts were numbered in order of when they should be opened, so unwrapping could only happen with all the kids together, and no one could ever claim to have more or less than anyone else.

Even on your birthday she would get presents for the other children, just so they had something to unwrap too. She knew that so much of the joy comes from just unwrapping your presents, so anything she could do to add just a little bit more ceremony and time only served to heighten our happiness by drawing it all out just that little bit longer.

Everything was wrapped – big and small – she even wrapped up all the sweets and chocolates. She must have spent an absolute fortune on wrapping paper, but there it was every year, Nan's sweety box (not jar!) piled high with sweets and chocolate, each individually wrapped in their normal wrappers, and then wrapped again in glistening Christmas paper.

Perhaps it all sounds over the top, it probably is, but I think there is definitely something about growing up with nothing, where one book (or one video tape of *The Lion King*) meant so much, that when you do have the means to spoil your family a little, you properly go for it. No one ever forgets how hard it has been, but if, just on occasions, you can make things that bit more special, to give the things you could never have had, then I can't see what's wrong with being a bit indulgent. I will never ever forget those magical Christmases as a child. They were some of the best days of our lives.

My nan helped me so much in becoming the woman I am today. She did everything she could to build on this little germ of an idea that I had to one day work with animals. She took me to every community and city farm in the North-west, multiple times, before we graduated to trips to Chester Zoo and Know-sley Safari Park, where I was thrilled to discover you could get closer to the animals than ever before, as the monkeys in the drive-through enclosure started to dismantle our family car.

She paid for me and my sisters to swim with dolphins in Florida, and paid again when we went whale-watching in Canada. If anything with animals showed up on television she would call me up right away, or videotape it for me to watch later. I was forever getting these videos with 'Farm Animals, Hannah' or 'Dolphins, Hannah' written in her beautiful handwriting right across the front. She ensured we could take the collies regularly to train in 'flyball', a dog sport where dogs race each other over hurdles, then press a spring-loaded pad to release a tennis ball, which they catch and bring back to their owners. I was forever videotaping our dog's performances for her to watch and scrutinise later on. I'm pretty sure she even found the clos-est flyball to us, just so I could get started in the sport, and dog training. She was definitely very encouraging about us getting a third collie at the time too.

She could see for herself just how much good spending more time with even more animals was doing me; in that way, I hope it did her good too. It was the same for all us sisters. Holly was into Disney princesses and doctors, so Nan ensured she was forever walking around dressed in a billowing princess dress, with a doctor's uniform pulled seriously over the top, and Caitlin was always getting anything to do with building or construction, so she received endless Meccano, Duplo and Lego sets. Holly *is* actually a doctor today, and Caitlin is studying for a PhD in biological engineering. Nan had spotted something in us early and, alongside Mum and Dad's support, made sure all of us would have the tools we needed to develop our passions, ultimately, into our careers.

Her unshakeable faith in me and my abilities made me believe that anything was possible for this little girl from the Wirral.

'Your belief guides my dreams.' I had those words tattooed to my left foot on my eighteenth birthday and still make sure it is the first foot into my welly every single time.

Chapter Three
Learning to Lose

When I turned 12, the carefree and communal St Andrews Primary School was swapped out for Wirral Grammar School for Girls.

The change was massive. I knew the rules before I'd even set foot in the door at the Grammar. The appropriate skirt length, the correct tie to couple with the traditional so-scratchy-it-might-be-made-from-Brillo-pads jumper, the ankle-high white socks that had to be worn in summer, the starched blue shirt, the dark-coloured bag, the cardboard blazer. I don't think I had ever worn a skirt, let alone one at the 'appropriate length', by the time I started secondary school, and that uniform was emblematic of how I generally felt most of the time at Wirral Grammar: a bit boxed in, like a sheep getting slowly squeezed with the rest of the flock into a one-size-fits-all pen; regardless of your species, personality, size, or whatever.

It could scarcely have been more different to my primary school. There, my mixed year group had all got on exceptionally well. There were no real 'rules' beyond obvious ones like 'Don't do things that could kill you or your friends', or 'Please try not to bring your dog to school'. I could see my home from the playground at St Andrews, and the atmosphere there was relaxed, gentle and inclusive.

Suddenly, all my friends from primary school, bar one, were

gone, and I felt thrust into this rigid and regimented all-girls school that had already removed a large portion of normal society through its 11-plus entrance exam. The school's top brass never seemed to tire of reminding us of this obvious fact – 'Remember, girls, you are in the top 30 per cent of the entire country' – but I didn't really understand why this was considered such a triumph of Wirral Grammar for Girls. If you have to pass a mathematics, verbal-reasoning, writing and English language test before you can even set foot on the premises, you are automatically guaranteeing an above-average set of students. It wasn't due to some magical fairy dust that sprinkled down from those red-brick walls.

Whatever their intentions were with these statements, it certainly didn't motivate me to work any harder. From the day I arrived, I felt we were all being groomed towards the ultimate pressure cooker: the final exams, the results of which would be the definitive yardstick for just how well the school was really doing.

It wasn't enough to simply pass your exams. Wirral Grammar needed straight-A students: Oxbridge candidates, vets, doctors, lawyers, engineers and political leaders (basically, they needed my sisters, and not me). In the Sixth Form they even introduced an extra level of segregation where that 'special' brand of person would be invited along to meetings to discuss their 'development' and to sharpen their claws for the tough entrance interviews that lay between them and admission to the next level of the elite world, whereas the rest of us were sent back to being just ordinarily privileged within this extraordinarily elitist system.

I prefer to just live in the moment and hated the pressure of thinking about those exams, especially as, in my mind, they existed on the other side of some far-off horizon many years in an extremely distant future. But pretty much every week, without fail, we were reminded that they not only existed, but we were expected to excel.

That culture, or ideology, or ethos, or whatever they called it, served to suggest to us that some of the divisions in society were necessary, and some were even desirable. A pretty bad idea in itself, but an especially poor judgement in an all-girls school, which, you should know, is the natural breeding ground for the sharpest form of bitchiness in the whole history of humans.

I was really lucky. I made friends easily across all the groups and cliques that, quickly and inevitably, developed at school. I think, to a large degree, the fact that I never got involved with any backbiting or name-calling meant that I was given some sort of diplomatic immunity at Wirral Grammar. It also meant that the victims of the bullying that did happen would feel comfortable enough to reach out to me.

Instead of just saying, 'Oh look, so-and-so has cut herself up again,' behind someone's back, I would go up to that person and ask if they were okay. That shouldn't be especially radical behaviour, but in schools like mine it was, and it meant I was probably overexposed to the depths of the unhappiness among some of my schoolfriends. I don't think my school was any worse than your run-of-the-mill all-girls grammar school; no one was getting physically bullied at least, but self-harm, eating disorders and serious suicidal thoughts were certainly not as rare as you might think. I probably shouldered a lot more responsibility back then, as a young and inexperienced woman, than I should have.

I still don't understand why some groups of women slag each other off, especially given how many barriers are already in front of us in our lives anyway, but two-faced people of either gender have always been a real problem for me. For better or worse, I try to wear my heart on my sleeve. If someone has pissed me off, you'd better believe they're going to hear about it face to face. I think it is one of the reasons why I love animals, and especially dogs. They don't tell lies, are steadfastly loyal, and what you see is always what you get.

My own school problems really came with the rules and never-ending academic pressure. I was never conventionally naughty, I didn't have a single detention or get into any real trouble, but I didn't just conform either. The headmistress didn't really know how to place me. I didn't fit into a convenient box: I was an outlier, drifting around somewhere between 'good' and 'bad'.

I really struggled to find the motivation to care about indoor learning on subjects that I didn't think were likely to ever be that useful in my life. Ancient history, quadratic equations, pi you can't eat, general-fucking-studies – surely the most pointless subject in the entire history of Western education – all of those things that if you have a computer you can discover for yourself within seconds, but for some reason we were forced to remember in our heads, regurgitate on a piece of paper, then immediately forget about for the rest of time.

I now know that I learn through actually doing things, especially with my hands (I can write as well with my left hand as I can with my right, by the way), and that it actually isn't that I can't take pressure or that I'm thick. Successful farmers must develop an understanding of the needs of multiple individual animal species, as well as the needs of multiple animals as individuals. Throw on top of that the challenges you have to surmount through the massively varied seasons, and the essential skills of time management, prioritisation and all of the logistical hurdles that come with running a farm efficiently, and I think it's fair to say that farmers are actually very capable of retaining highly complex thoughts and ideas, under conditions that are often extremely stressful.

I prefer to be judged by that practical environment over a long period of time, rather than chucked into the confines of a stifling and silent exam room with just a couple of hours to demonstrate knowledge that's probably taken years to learn. Yet that's the reality of how we choose to test nearly all young people today, and if you can't perform in that single, minuscule and narrow-minded moment, then you wind up feeling like you've just completely failed as a human being.

I know I am far from alone in thinking this way, and I also know that people with skills and thought processes like mine regularly fall right out of mainstream education and, later, sometimes even life. It's doubly tragic. They are denied any opportunity they might've had to fulfil their potential, and the world is denied, quite possibly, extraordinary, revolutionary, skills and thoughts that you just can't get from conventional people.

The inflexibility I encountered at my school was my major problem with the system there. At Wirral Grammar, and many other schools of that ilk up and down the country, I should imagine, the pressure to maintain their standards and standing is so intense that if the status quo is working then why would you ever dream of taking any risks? Well, my answer to that would be because of what an education system should ideally be: focused on what's best for the students, inclusive, and absolutely uncompromising in its ambitions to reach every child.

You might argue that it's impossible to design a whole education system around students like us, that we would end up with some multi-headed beast of an education system and a scattergun way of testing to suit the whims of every single child and their specific individual needs. That it would be impossible to teach, and unfeasibly expensive. But I just don't believe that I'm in such an extreme minority that absolutely nothing should be done either. In my profession I meet hundreds of people that think like me, and if you sidestep into all the other professions that are similar, I absolutely guarantee you will find hundreds of thousands, if not millions, more. I'm not asking for a complete radical overhaul of the entire education system, just that more attention could be paid to all the children that do think differently, instead of just assuming we are all stupid or bone-bloody-idle.

I'm certainly not qualified to start making specific suggestions on exactly what we should be doing differently in our schools, but if you are a young person reading this, and it resonates, then you are always welcome on my farm to try something else.

I got through my GCSEs unscathed, but that was when the academic side of things took a major upswing. Back-to-back exams for two years straight, before I could even think of university and an escape plan into a career in creatures.

Given my gut feelings about the academic system by the time I turned 16, this was highly unlikely to be an easy ride for me. I needed that secure base of my stable home more than ever, and that was when we lost Nan.

She was only 54 when she fell.

How many times must she have safely gone up and down those steps? And how many times does anyone really ever hurt themselves, I mean properly hurt themselves, just falling down a set of household stairs?

It was about seven o'clock on a cold night in November when my parents had a phone call to say our nan had had an accident. My Auntie Sarah came straight over to stay with my sisters and me, and Mum and Dad shot round to her house right away. By the time they arrived the ambulance crew were already there and it was obvious she was in a very bad state. She had fallen carrying food. It was all over her, up the walls, across the stair carpet. Nan was unconscious.

For a few years we had been taking round our daft-as-a-brush collie dog, Rosie, just for a bit of daytime company for her. Nan loved her so much we ended up just leaving her there, this fun-loving, free-spirit dog that ate absolutely everything in sight. There was Rosie that evening. Sat down next to my nan. Watching over her in a state of such intense concentration and deep concern that not a single scrap of the food had been touched.

Dogs know when something is wrong, but that was my first insight into just how powerfully attuned to us dogs, and especially collies, really are. In the immediate aftermath of that awful accident, that Rosie had probably witnessed, she knew, before all of us, that our Nan was not coming back. Rosie had already begun to grieve.

Nan was blue-lighted to hospital as quickly as possible, but her pupils began to heavily dilate in the back of the ambulance. The scan of her head was completely white down one side. The bleed on her brain was massive, like she'd been in a high-speed motorcycle accident, not some trivial domestic tumble. It was catastrophic and utterly unsurvivable. That was it. Nan had gone.

*

The shock I went into the next morning, when my mum sat on the end of her bed and told all of us that Nan was dead, was absolutely and instantly numbing. It felt like my whole world had ended.

I didn't speak for three days. The funeral was a blur. I blocked it all out. Christmas was awful. Nan had already wrapped most of her presents before she died in November, such was the scale of planning required to properly execute her Christmas campaigns. I can remember every single present we opened that had been so lovingly wrapped by her, dreading the inevitable moment when we all reached the very last gift that she would ever have been able to give. The wave of emotional heartbreak was so all-consuming, it was almost suffocating.

I know a lot of people might find it hard to understand why the passing of my nan hit me so hard. So many of us have experienced the unique love of a grandparent, and so many of us have seen them pass, and coped. Death is inevitable and grandparents are supposed to die before their children and grandchildren, so while you are supposed to be sad, it is also expected that you are likely to see them out of this world. It is a fundamental fact of life that no one lives forever.

But this was so sudden and unexpected. Nan was just 17 when she had my mum, and my mum was just 20 when she had me. My nan was young enough to have easily been my actual mum, and in many wonderful ways she truly was like a second mum to me. Not that I didn't get everything I ever needed from my mum already, but having a nan like mine was an additional blessing. Both Nan and Mum went through so many struggles as young mothers, but the major benefit of what they did was that we, as grandchildren, were so much more likely to get so many more decades with them than most could ever dream of. That was where so much pleasure, richly deserved, was going to come for my nan: our graduations, our weddings, the births of our children, and maybe even the births of our children's children.

All of that was stolen from us when she simply fell down the stairs carrying a plate of food, and that robbery, of the special time she had absolutely earned, never feels more raw than in those moments where you look around and expect to still see all your loved ones standing right there beside you.

Things were significantly worsened by the behaviour of a man I have not yet mentioned. Nan didn't meet Jim until she was 31. They married three years later in 1989 when Mum was already 17. I grew up believing Jim was my grandad, and there was zero distinction between the way I saw my nan and the way I saw him. Perhaps he was not the father-figure to my mum that Bill was, but Mum loved him regardless, and my dad made sure he asked both Jim and Bill for my mum's hand in marriage.

Jim did everything with us, just like my nan. In fact, when my nan couldn't be there it was Jim who came on the holidays, Jim who came to the family gatherings, and Jim who would take us out shopping nearly every weekend. He even worked with my mum and dad. He was absolutely and deeply woven into the core fabric of my entire family. The perfect example of everything a grandad should be.

Looking back now, there were signs something was afoot immediately after Nan died. He came and lived with us until we had the funeral, and commented to my mum that he felt we, the grandchildren, were now treating him differently. I certainly don't think that was true.

He became increasingly distant after the funeral, to the point that we weren't really seeing him at all, which was extremely weird given how extraordinarily close we had all been for the entirety of my life to that point. Not long after this, my mum sat me down and explained Jim wasn't actually my biological grandad. I was really shocked, which was, in hindsight, perhaps a little bit daft, especially given my younger, and much more logical, sister Holly did point out she had been suspicious as to how he could possibly be Mum's father when our dad had taken all the photos at his wedding with Nan. I can laugh at my naivety on that one, but it still shouldn't have made a blind bit of difference. It certainly didn't in my mind. Jim *was* family.

Pretty soon after that revelation Jim completely severed ties with us. I just could not get my head around how someone who was so involved, and so loved, could just suddenly up-sticks and abandon us.

It came out soon enough. As Nan's will became available it became clear that everything had been signed over to Jim. Nan had managed to get a significant sum of compensation for her illness from the company in the end, but it hadn't been touched. That was what she had wanted to pass down to us.

She had said it enough times to my dad, who she dearly loved too: 'No matter what happens, Stu, everything I have is going to my girls.' But ultimately it was Jim who got all the money, and pretty much everything else, when Nan died. He let my mum back into my nan's house to get her wedding dress, at a time when he ensured he was not going to be present, and that was it. He even got to keep Rosie the Collie.

Jim's relationship with Nan couldn't have just been about money. There was no money to be had in the beginning, and even so, you can't fake the sort of love they'd had for over twenty years. I *still* genuinely believed he had adored my nan, and I *still* believe he had loved us too. What he did was, and is, inexplicable. He, or at least the person he was to me, also died that day: on the stairs, right alongside Nan.

Jim maintained a relationship with Bill and Betty, Nan's parents, for a while, giving me some hope that things might get sorted out. I wrote him a letter, that I could never quite send, and for a few years I clung on to the belief that he would eventually come back, and that all might be forgiven. My eighteenth birthday ticked past with nothing, even on my twenty-first I was still thinking he might just send a card, or something, anything, as a gesture to say that there was still some love and caring deep down in there, even after all this mess.

But the card never came and a reconciliation never happened.

Anger has since replaced all the sadness and confusion. My nan had no fault in her life's story, but Jim actually chose to leave us, and because of that, as much as I would like answers to all my questions, I actually don't feel like I could give him the time of day. It was the first time in my life I had ever felt truly betrayed by someone, and the cut in my heart that opened when my nan died was absolutely torn apart by that man.

I can never forgive him.

Up to that point there had never been a moment where there had been anything other than happiness in our home. I had been very lucky that I hadn't lost anyone so close in my life before that point, but that was no real comfort to me back then.

For some time, everyone in our house was deeply sad. It had never been so quiet, but I wasn't going to be the one to break the silence and actually talk.

Betrayal and bereavement are a truly toxic combination,

particularly when you throw them on top of all the normal teenage anxieties and insecurities pretty much everyone goes through in their last couple of years at school. The fact that I was also cultivating quite serious issues with conformity, and the whole system at my school, definitely didn't help either. I just didn't have the maturity to reset myself and go again, but what 16-year-old would have? My response wasn't some dramatic booze-fuelled breakdown into antisocial daftness, though; I just shut down completely.

I was a teenager who was into everything, but my life shrank almost overnight into a comfort-blanket routine of chicken fried rice on a Friday evening, TV, a handful of friends, and MSN Messenger. I didn't go out that much. I certainly didn't do my homework. I didn't really do much of anything at all really, I was just vacant. Sleepwalking through week after week of my life.

There was this children's television programme that was on when I was a kid called *Bernard's Watch*. Bernard was this very ordinary young lad that either wore a jumper that looked like it had been knitted by his mum, or an absolutely awful mustard-coloured sports jacket, but he also happened to carry a magical pocket watch that he could use to stop, or even rewind, time.

There were two key rules to using the watch: you couldn't use it to commit crimes, and you couldn't use it to hurt anyone. So, Bernard used it instead to do lots of wholesome things, like helping the elderly safely cross a road, or swapping salad for chips for a veg-dodging mate, or bringing his grandad his glasses so he could compete, and later cheat, in a charity darts exhibition match. Okay, so not always all that wholesome, but in every episode, there was always this bit when Bernard would click the pocket watch and everything and everyone would freeze. Time would stand completely still and Bernard would wander through the scene while the rest of the world and everyone in it, basically, completely ceased to exist.

That was exactly what my grief felt like. I didn't know how to get out of it, and I didn't want to either.

I had my first series of AS exams coming up in the January after Nan died and Jim left. My mum tried to persuade the school to let me defer them until later in the year, but the headmistress was insistent. I took them. And failed every single subject.

I didn't do a whole lot better in the summer. I mean, I couldn't have done any worse. I had double exams: retakes for everything I had failed in January, and all the new ones for the end of the academic year on top. I scraped through, but I was so far off what I was actually capable of, it felt like I really shouldn't have bothered in the first place.

People grieve differently, just as they learn and study differently. I wasn't asking for a huge amount, and I certainly wouldn't have deferred for a whole year – my nan would've been absolutely fuming if I had – but I really needed some more time.

The turnaround, when it came, didn't happen in one dramatic Hollywood 'and now I see the light' revelation. Even as my parents and teachers tried to offer a hand and show me how close I was to actually throwing my dream to work professionally with animals directly into the dustbin, I still couldn't quite pull my head out of my arse. I always live in the moment, but right then, I was seriously trapped in it. But, bit by bit, things slowly came together with a series of tiny wins.

The anniversary of Nan's passing was coming up and there was no question that she would've been really sad to know how badly things had been going for me at school. Year markers, be they birthdays, anniversaries or Christmases, are always really important times of reflection in my house, and this was going to be the biggest moment of all. I had a little word with myself. I couldn't let this carry on into a second year. I began to realise I was letting my nan down.

I started dyeing my hair red. Not a big deal for an average teenage girl, but it was for me. I've already spoken, in the opening chapter of this book, about what having red hair means to me. The first time I did it, though, it was also about trying to find a way to move on by just looking a little differently at myself. The red was especially significant because back in the final year of primary school we'd had something of a prom. It was all a bit silly – 11-year-old girls getting dressed up in big dresses with the boys wearing black-tie and waistcoats – but my nan had got me ready for that night and helped me dye the tips of my hair red, tying it all up into one big messy bun right before I left for the disco that evening. That was going to be the only coming-of-age event she was ever going to be part of, and I wanted to remember that too, somehow.

Sometimes, I realised, you do actually need to look back a little, to start looking forward.

I got a weekend job at Knowsley Safari Park, working with the very sea lions I had seen so many times as a child on all those trips there with my nan. The memories of that place were very much ones I had made with her but, without question, being there and doing something much greater than just visiting was exactly the sort of thing she would've hoped for me. I also started working with my uncle, helping him out with bat and newt surveys in old decrepit buildings scheduled for demolition and development. That was a terrifying job, taking me in the dark to places like an abandoned lunatic asylum and a dilapidated theme park, long since closed and sucked of all its fun. At one point a man, peeping at me through a fence, nearly caused me to drop dead from absolutely shitting my pants. I will never forgive you, Uncle Fran, but both jobs, at the safari park and the surveys, solidified in my mind that I definitely wanted to work with animals professionally and academically. It also meant that I could put together a quality personal statement, based on hands-on experience, for my university application. Still, I knew I definitely couldn't keep failing my exams and essays.

When one of my teachers asked me to see them after class shortly after I had submitted an assignment I immediately thought, *Here we go again, more trouble.* But no: 'Hannah, that piece of work you have just done was the best piece of work you've done all year. Keep that up and you are going to be fine.' That was just what I needed. That small boost, praise for a single piece of work, was enough to keep me focused on handing in essays with actual effort and considered thought for the rest of my time at school.

I realised then that I didn't have to agree with the system, but it was a stepping stone that I would *have* to cross if I was ever going to get to the next point in my life. The school might need the A star students filing towards Oxbridge, but I just had to focus on the grades I needed to go to Liverpool John Moores to study a BSc in Animal Behaviour.

Another big moment came soon after that. I was tested for dyslexia due to the fact that I just couldn't physically write that fast. My handwriting is incredibly neat, maybe it's me focusing too much on handing in essays that look like a piece of art, instead of concentrating on their academic content. At any rate, I did not have dyslexia, but I did have enough of a problem for me to be granted extra time in exams, and a computer to type all my answers onto. It also meant that when it came to my final exams I was happily exiled from the rest of my classmates, tucked away in my own room and away from the mass anxiety of the examination hall.

As an adult I can now see that my struggles were not wholly due to the system of education at Wirral Grammar, and that my parents were absolutely right to send me there. Who knows how I would have performed in another environment? Especially given the very limited choices elsewhere in the area. I could easily have bombed out completely. There is also no doubt in my mind that there were actually many good teachers at my school, who had tried to understand my needs and pull for me behind the scenes. Some had tried to intervene immediately after we lost Nan and, certainly once I had found my way back into the room academically, many more went out of their way to help further. That laptop, the extra time and the room away from the exam hall were simple, but absolutely game-changing. It does show that small concessions really can make a world of difference to students like me. In spite of everything, I can now look back on that time and appreciate just how lucky I was. I am extremely grateful to have been given another opportunity to achieve and, as challenging as it sometimes was, going to Wirral Grammar was undoubtedly a rare privilege that hundreds of thousands of young people, and their parents, would walk through fire to have had.

I logged on to the UCAS A-level results website long before I was due to pick up my results in person. There they were, the letters I needed. Everything that had seemed so far away, for so long, was suddenly right there on the screen, in black and white.

I burst into my parents' bedroom in floods of happy tears. We had turned a corner, and finally, I was on my way.

*

One of the hardest things about losing my nan so suddenly was that we never got to say goodbye.

If I could wish for anything, I would ask to hold her hand just one more time so I could tell her just how grateful I am for everything she did, and that all her efforts to keep all of her grandchildren's dreams alive were never wasted. I know my nan would be so proud of all her grandchildren today. She put an incredible amount of effort into all of us, but even if it hadn't all worked out in the end, she would've loved us all anyway, without condition. That sort of rare love was her greatest gift.

Thank you, Nan. Thank you for everything.

Unbeknown to me, when I was born Nan had put money into a bank account under my name. Mum was a joint executor of this 'secret account' and, a few years after Nan died, Mum released the money to help buy my first proper thoroughbred sheepdog.

That young dog, presented to me for my twenty-first birthday, was Fraser. A piece of my nan's spirit to flow by my side from that day forward.

When I am with him, I still feel her.

Chapter Four
Larry the Lamb

'I think something is happening, with that sheep, right up there.'

Way up in a corner, away from all the other sheep in the flock, I had spotted a single white ewe sheltering herself up against a dry-stone wall. There was no bad weather or any sense of threat in the rest of that field, she had just taken herself quietly away, and I didn't yet know why.

My whole family were on holiday in Coniston, a picturesque Lake District village sat at the northern edge of Coniston Water, a five-mile-long lake pooled up at the base of a deep glacial trough.

We were on a walk that we had repeated many dozen times as a family before: past the steam-boat jetty and Coniston's bicycle hire and cafes, beyond the lake's pebble beaches, pretty islands and the mixed woodlands that inspired Arthur Ransome to write *Swallows and Amazons*, up and away onto a pathway that leads along the edge of a sheep farm and its fields, before it pulls away into its eventual destination: the dark peaks of the Furness Fells.

We all slowly approached the sheep, quietly following that footpath as it ran alongside the dry-stone wall that provided the ewe with her sanctuary. Eventually we were within four feet of her. We gripped the ancient hand-placed rock on the wall's top and peeped over.

'I think she's having a baby!' my sister Holly whispered excitedly. The ewe's belly was improbably swollen and, clearly, she was starting to strain. If we didn't all actually hold our breath in that moment, we certainly fell absolutely silent. Something magical was about to happen.

I now know that ewes will often take themselves away to give birth. The instinctual feeling of vulnerability as labour begins demands shelter, safety and quiet. Well away from potential predators or anything that might seek to interfere with the intense focus the birth rhythms demand. This is a common habit shared among most live-bearing animals, but pregnant ewes do have some distinctions in these precious moments too. A series of hormonal changes occur in the lead-up to this point, naturally guided and initiated by the lamb. The ewe may have stopped eating, and her udder and teats will have begun to swell with her first milk. In sheep, like humans, this early milk is called colostrum. Like ours, it is filled with protein and antibodies, but the ewe's colostrum also carries huge levels of fat, probably due to the fact that a newborn lamb will be up on its feet within minutes of birth.

Her vulva dilates in readiness to deliver and soon her whole rump section begins to appear a little shrunken. The muscles there are beginning to relax behind the lamb, as it descends through the womb. Further down the uterus, the linings and muscle fibres tense up in contraction, and the ewe herself will often appear to be in a slightly hypnotic state. We call this faraway look 'star gazing', and you could be forgiven for thinking the ewe was a bit spaced-out and not really all there, but actually the exact opposite is true. The early stages of labour, when the whole focus of the ewe's body is on pushing the lamb towards its eternal external air, require such phenomenal levels of mental concentration that in that moment the physical world outside her mind's-eye almost ceases to exist. It is another reason why that early search for a safe-feeling shelter is so important, and it is certainly why we were able to watch on without disturbing the natural course of nature that day.

Our ewe suddenly rose to her feet, walked in a neat circle, and then lay back down. Often the ewe will reposition herself several times during labour, just to find the most comfortable position, or to motivate the progress of the birth through a little physical movement. The ewe in front of us did it just once, and then things really started to happen.

The ewe began to talk in a special birthing language that you only really hear between mums and their lambs. It is not a conventional 'baa', it is so much more subtle and discreet. It's like a low 'hum', a muttering noise that is absolutely unique to her and her child. Even as the lamb grows, a single call using this tone will still extract the mother's lamb from flocks several hundred strong.

You get some very dramatic ewes during birth. Sheep that will go absolutely mental with screaming noises and emotional cries during labour, all indications that they absolutely believe they are about to die. But our ewe just gave one more deeply composed and prolonged heave, and suddenly, we could see the lamb's two front feet and nose presenting at the opening of her vulva. All three of those tiny body parts were compressed together in a neat little triangle. I couldn't have known then, but it was the perfect birthing position.

One more push brought the whole lamb sliding from its mum in a rush of pure-white fleece, blood and fluid from the now-burst amniotic sack that it had been growing in during its life inside.

I had never seen anything like it. The whole family were in a state of dumbstruck awe throughout the birth. It is not very often that every member of the Jackson family is rendered speechless, but there we were, entranced by this incredible moment. As soon as the lamb was out, though, the spell was broken:

'Hell fire! Did you see its nose sticking out?' shouted Caitlin.

'Look at all the mucus on it!' cried Holly.

'I think it's about to stand!' observed Dad.

'Well done, Mum, wow, she looks tired!' said my mum.

But I was still stood silently, held within my own private 'star gazing' moment where all that mattered was just that ewe, that lamb, and that whole new life, that I had witnessed tumbling dramatically into the world.

The ewe continued to act along the ancient and eternal maternal codes that are embedded deep in its subconscious. I found it deeply meditative to witness. It was a feeling that made me feel rooted to the earth, as well as that spot, in a way I just hadn't ever experienced before. I didn't know much, but I did know I'd just seen something that had happened between mammals for millions of years.

The reassuring muttering noises continued to flow from the ewe as she cleaned the yellow birthing fluid from her lamb. She extended her thick purple tongue and ran it over the lamb's nose, before heading down the length of its body, diligently working until all the mucus – from mouth to belly button to bumhole – had been removed completely.

The freshly cleaned snow-white lamb rose onto its fragile-looking legs and began to wobble and totter its way towards the underside of its mum. Its instincts were powering it forward now too. It was about to take its very first drink of its mother's milk, right after its first steps, and its first breath of fresh air into its lungs, all within minutes of its birth.

It was such an intense rush of so much new information that morning. I realised I had never seen an animal born until that day, in fact, I hadn't even seen the tongue of an adult sheep before. My closest interaction with a young lamb before that was one that was actually several days old, that we had paid to feed with a bottle during one of many urban farm visits with Nan.

All of the cats and dogs we had welcomed as pets into our house so far had arrived as kittens and puppies. Your average pet owners, even ones like me that have had so many, are nearly always completely removed from the blood and gore of the actual birth and, more often than not, when your cat or dog reaches sexual maturity, then it's packed off to the vets to get spayed or neutered. Polite words for the actual removal of an animal's reproductive system.

The dirt, distress, even, at times, quite violent scene of a ewe giving birth either puts people off anything to do with sheep for life, or they just intuitively get it. There was something about that process that just spoke to me that morning. I have thought about that lamb so many times I even gave him a name: 'Larry'.

What if we hadn't done that walk that day? What if the labour had been horrendous and ended in catastrophe? What if we had walked past the field and seen nothing? I believe in fate, though, and I think you ignore it at your peril.

Larry's birth was so visceral, so primeval, and so awe-strikingly real, it was impossible to forget. I snapped out of my trance with a thousand burning questions that I naturally wanted answering immediately. Fortunately for me, within the hour I had cornered a poor farmer. Unbeknown to me at the time, my journey towards a completely new career had just been kick-started.

Before Larry, my whole life had been heading happily towards the study of whale behaviour, specifically, the highly intelligent, charismatic and fairly ruthless cetacean known as the orca, or killer whale. Between the second and third year of my under-graduate degree in Animal Behaviour, I undertook a scientific placement as a volunteer whale-watcher.

It hadn't been a straight journey. I briefly thought about becoming a vet, picked up some work experience at a local clinic, but found it all too samey, too sad and, probably, too much studying. The day I knew it definitely was not for me was when a dog was rushed in with blood spilling from every single hole in its body. It had just been hit by a car and we did absolutely everything we could to save its life. The upset I felt, zipping it up inside its body bag later that day, was so devastating I had to resign myself to leaving that career path well behind.

I had briefly trained sea lions at Knowsley Safari Park and had even entertained childhood dreams of emulating the little boy in *Free Willy* and one day working with orcas at Sea World. That died the moment I first saw orcas in the wild while on a family holiday in Canada. Seeing that pod instantly reduced me to tears and, long before the revelations we have today about the unethical practice of keeping orcas captive, I intuitively knew that keeping any wild animal purely for entertainment was absolutely wrong.

I worked hard to find anywhere I could study orcas on a placement, and eventually found a research project based right over in the remote far west of Canada. A successful Skype interview led to a long-haul flight to Vancouver, a hop onto Vancouver Island, a bumpy car journey with a bloke I'd never met, two ferries, an inflatable rib boat and, several days later, I had successfully arrived for work.

I was stationed on an uninhabited island on the wild western Pacific coastline of British Columbia. West Cracroft was everything you might imagine when you think of the Pacific Northwest's true wilderness. Smuggled among a maze of rugged islands, it appears like it has been smeared on the map by the fine head of an artist's paint brush. A three-mile-wide finger of water divides West Cracroft from the much larger landmass of Vancouver Island, and this narrow channel, the Johnstone Strait, funnels the super-rich waters of the North Pacific into a densely concentrated soup of krill and micro-organisms. The Johnstone Strait attracts fish and aquatic wildlife from truly vast distances to hunt and eat. The humpback whales found here were regularly logged to have travelled over 3,000 miles, in one of the longest animal migrations on earth; but our patch was just one of many feeding grounds hemmed in along a vast sweep of meandering coastline that stretched some 20,000 miles along the length of British Columbia and all the way round to Alaska.

The Johnstone Strait was a natural super-highway for an array of whale species. Lining up alongside the humpbacks were fin whales, grey whales, sperm whales, minke whales and, top of the tree – for me at least – were the orcas. And West Cracroft, with its clear mountain summit vista overlooking the strait, and its natural harbour for launching research vessels, was *the* perfect place to study them.

When I say the island was uninhabited, it wasn't just that there were no people living there, there were no signs of any human habitation at all. Our basecamp consisted of little more than a scattering of wooden pallets along the water's edge to pitch our tents. I picked a pallet that was on the edge of the camp, just for a bit more privacy, but I still made sure there was another tent positioned between me and the island's densely forested interior. Selfishly, I figured their tent would be attacked by bears before mine, so I might get a bit more warning before I was inevitably gobbled up too.

West Cracroft was only about 15 miles across, but the side of the island opposite from the Johnstone Strait was only one simple 100-metre low-tide hop-skip-and-jump into the entire hinterland of British Columbia. That area of extreme wilderness is three times the size of the British Isles, with less than a tenth of its human population, and a higher density of bears than anywhere else on earth.

The fur-covered and sharp-toothed population of the island didn't just extend to the bears. We were also warned about cougars, wolverines and wolves, so a predatory encounter felt pretty much guaranteed at some point. In fact, a brown bear had even been spotted in our camp right before we'd arrived, depositing its big fresh poo on the floor between two pallets that no one was rushing to occupy. I was no longer in the Wirral, and this was no petting zoo, but I dearly wished Nan could've seen me then.

The risks of dying some grisly death at the paws of a bear, or any other North American predator, are actually very slim. Bears only really attack when they are surprised, so I made sure that whenever I was on my own on the island, even if I was just going to the toilet, I would have constant conversations with myself and those bears. 'I'm here, bears, it's only me, horrible-tasting Hannah, walking to the toilet. No need to get startled and end up eating me.' I would always make sure I slept in the centre of my tent too, as if the nano-second of extra time it would take for a bear, having already ripped apart the flimsy canvas wall, to reach out from the edge of my tent and onto the little red-haired sausage roll in the middle would've made any difference whatsoever.

By the time I came to leave, some five weeks later, I'd actually only had two experiences with the island predators. One was iconic, and absolutely spine-tingling – a large solo male wolf howling out into the darkness while we were sat round the camp-fire one night – and the other was a pretty undramatic encounter with a big brown bear. I say 'undramatic' as it was happily at the end of my long lens, plodding its way along a beach many miles from me, my tent or my unusual toilet habits.

The truth I learned on West Cracroft is that the presence of apex predators, in good numbers, is the true indication of a healthy natural environment. We weren't likely to be eaten because we were in a place where wild food of all types was in abundance; it was about as close to a pristine wilderness ecosystem as I could ever expect to experience.

I was stationed at West Cracroft with half a dozen other researchers from all sorts of backgrounds. Our job was to monitor just how increasing boat traffic in the channel was affecting the natural behaviour of the whales, which meant spending long days studying their everyday lives from either the top of West Cracroft's exposed mountain peak, where we had an established viewing point called the 'Eagle's Eye', or from the seas in a research vessel or kayak.

Soon, I was able to identify the individuals in the local orca population purely on the sight of their dorsal fins. All orcas have their own distinct fin shape and pattern, but the orcas in this area were special anyway. They were among the only pods in the world that engaged in a practice called 'beach-rubbing', where families of orca would swim towards shallow shingle beaches and rub their bodies all over the stones on the sea bed, often within only a few metres of the tide line, and with just a skin of sea water covering their immense bulks. It's hard to say whether they do it for any physical benefit, as orcas naturally shed their skins over time anyway, and this behaviour is so specific to this isolated group of orcas, that really the only plausible explanation from the experts was that it was a form of 'socialisation'; a fun, slightly reckless (they could easily end up beached) family activity that binds a group tightly together.

Without question, there are elements of orca behaviour, alongside other intelligent animals like dogs, that are comparable to people. Outwardly displaying joy or choosing to do things purely for the fun of it is certainly one, but I was always careful to not fall into the trap of believing orcas are just like people. The sheer brutality of nature in its rawest form is a serious step away from the anthropomorphism on display in the Disney films of my youth, and the orcas were almost uniquely cruel in their behaviour at times.

They chase and kill minke whales for sport, hunting them in packs for hours on end before eating just the tongue from the minke's corpse. Little arseholes. But not as bad as the true terrorists of the sea: the superficially cute and cuddly-looking sea otters, the subject of children's books and greetings cards the world over, which I also noted were taking great pleasure in raping the resident baby seals.

Orcas have their own unique ways of communicating within their pods. Quite similar to ewes and their lambs, it is a language they have developed specifically for their own kith and kin. They also respect a strict matriarchal hierarchy, that sees a solo dominant female at the head of the pack, who only leaves her role as leader in death. That can take an extraordinarily long time; there was a female boss of one pod in Canada, called 'Granny', that was estimated to be 105 years old.

Their behavioural intelligence extends well beyond their ability to communicate and choose activities that have more than mere survival at their heart. I've witnessed orcas expressing complex emotions and can never forget the day on West Cracroft where I saw an orca mum carrying her dead baby calf. It was just like a funeral procession, and there was no doubt in the way that the rest of the pod were swimming that there was a deferential and sombre tone among the entire group. The mum quietly carried that child for many days on end.

Like collies, orcas will show off in moments of great joy, too. One morning I watched a single male spend hours trying to hunt a large Pacific salmon. Herding it, funnelling it into corners, trying to trap it, just like my collies do when chasing down a rogue sheep, until eventually its giant jaws clamped shut and it finally had its prize. Instead of just eating it, though, which would be the action of an animal whose behaviours are limited only to hunting and killing, this orca rose straight to the surface, leapt, and displayed its triumph held within its jaws. That orca was trumpeting its achievements by first showing every living thing in his patch of sea just how good he thought he was at hunting, before eventually consuming his catch. There was no survival necessity for that display: the orca could easily have eaten it without doing anything, but it didn't, and the longer I spent, and the closer I got, the more I could just see in their eyes that there really was something else, something bigger, going on within their brains.

I didn't realise how much that intelligence, that emotional connection, and that sense of a mutual understanding was something I truly needed to find to deepen my love for the animals I chose to work with. I can look to my dogs now and feel it all the time, the sense that I can understand and recognise their behaviour and feelings – happy and sad, good and bad – but also that there is a two-way reciprocity in that: that my dogs can also look and understand what I am feeling too. It was something I could only ever feel in one direction while I was out in British Columbia. I could observe and identify the immense intelligence in the behaviours of the orcas, but I would forever be an observer, spying on their world from a distance, without them ever really sparing any thoughts or feelings for me.

Away from work, our time was spent very simply. I lived out of my small tent, washed with rainwater, if we had it spare, or I just jumped in the sea. I sourced occasional electricity from our solar panel, but mostly just did without, and I ate almost an exclusively vegan diet. There was no fridge, so no way of keeping food fresh that wasn't dried or in a tin. If we wanted fresh fish it had to be caught, usually by ourselves with fishing lines or, if we were very lucky, donated by the First Nation tribe, the Kwakwaka'wakw, the human custodians of the land we were working on.

It taught me a great lesson. I actually needed very little to be really happy as long as I was doing what I wanted with my life. It put me in great stead for later jobs where the accommodation was pretty basic (it could never be more basic than West Cracroft) and it put me in a mindset where, without any distractions, I could develop an even deeper appreciation of the natural world. I spent hours snorkelling and observing how all the purply-bruise-coloured sea urchins would suck themselves onto the rocks in gangs of hundreds, or how some of the sea birds could actually submerge, swim and hunt right beneath you as you floated around the gin-clear seawater. With no unnatural noise to speak of, I soon found myself becoming so attuned to the wild I could identify marine species at night purely from the sound of their blowholes. The humpbacks had a huge elongated blow, having returned from extreme depths; the porpoises had a super-fast explosive noise, like they had over-pumped their bike tyres; dolphins had more of a gentle 'poof' sound, like what you might get if you jumped on a bean bag a bit heavily; and the orcas, well, take that dolphin 'poof' and basically put it on steroids. Absolutely unmistakable. I would lie back in my tent and have to just pinch myself at my luck at having such an incredible opportunity to be out there with these magnificent beasts.

The vulnerability you feel in the proper wilds is almost like no other. It is deeply humbling to be wrapped up in your sleeping bag, knowing you are truly at the mercy of the wildlife and, really, in the scheme of things, actually quite a weak animal when you are stripped of the human technology and hardware that makes us feel so artificially superior. It is a good thing to experience every once in a while, the idea that there are many bigger and greater things on earth than us.

Canada was light years ahead of my home country in terms of its environmentalism. The wilderness areas and National Parks were proper conservation zones, with robust laws, enforcement and punishments for anyone seen to be breaking the rules. They were so aware of the need to recycle, and at least a decade ahead of where Britain was in terms of eradicating the use of single-use plastics and plastic bags. In places like West Cracroft it was impossible to not feel a deeper spiritual connection with the land, but in Canada in general, people just seemed to be more switched-on to the need to care for this planet, and much further down the line in thinking seriously about man's often irresponsible relationship with nature.

I was extremely grateful to all the people I worked with over the course of those weeks. They all helped me understand the world differently, but one man had more influence on me than any other.

Ernest was a member of the indigenous Kwakwaka'wakw people. He was fair-haired and friendly, absolutely the joker of the group, but he was also deeply passionate about his heritage and had a hand-etched Kwakwaka'wakw tribal tattoo ringing his belly button.

He would become intensely serious whenever he talked about the lore of his land. The Kwakwaka'wakw see a deep emotional and spiritual connection between man, the animals and the landscape. Ernest opened my eyes to their way of thinking, and he taught me how it was possible to take away from nature without ruining it altogether.

One very clear memory I have with him was the afternoon we walked together to a grove of cedar trees. He wanted to show me how to make a traditional Kwakwaka'wakw bracelet from the cedar's bark, which sounded quite straightforward to my modern Western ear, but Ernest taught me how careful we had to be from the outset, even when it came to just picking the right tree.

The tree we chose had to be one which could cope with the small amount of damage we intended to inflict; once selected, Ernest went into real detail about how much width of bark the tree could withstand us harvesting, and how deep we could cut.

The focus was not only on the product we were making, it was on ensuring that we would not leave any long-lasting damage to the tree after we had taken the materials we needed. To Ernest, the natural legacy of the host was as important as the loan of its skin.

In Britain, that tree would probably have been chopped down to make it easier for the woodsman to make his cuts. Nature, for us, for so long, was only ever considered in terms of its monetary value; streamlining the process of taking from it, to make us more money, more quickly, was the only real priority. Things are slowly changing now, but in many ways the damage, especially to our ancient forests, is already irreparable.

We cut a long strip of cedar bark together and Ernest began to show me how to plait and weave the bark into a tight ring, which, when he had finished, was the finest piece of natural jewellery I have ever seen. I love it so much I hardly ever wear it, preferring to keep it tucked away within the ornately carved cedar box I use to store all my artefacts from that very special time.

'Hannah, when you take anything from earth, you must thank it at the same time,' remarked Ernest, as we were leaving the cedar grove. 'In doing so, it reminds us that this tree is as important as an individual, and that if we were to take too much from it today, and be too greedy, it can no longer live its life here any more.' He explained the bracelet would only offer protection to the wearer for as long as it stays intact, and that when it breaks, the tree can no longer keep you safe.

Ernest taught me the rudiments of trying to live a life symbiotically with the earth, instead of always just looking for ways to dominate it. That philosophy would change the way I viewed everything in my life, and later it changed my approach to farming too. It made me more aware of the need to try and raise my animals as harmoniously, and as happily, as I possibly could, even if I was to be working within a British landscape that was always going to be vastly different from the unfettered wilds of Canada. More than anything, I continually remind myself that the animals I rear for food today are living things and not just the emotionless end product of some heartless factory assembly line. Taking their lives should illicit humility and a sense of privilege or, at the very least, deep gratitude.

The whole experience in Canada had a profound impact on me. By the time the family holiday at Coniston came around, I was right at the end of my final year of my undergraduate degree and felt more certain than ever that I was going to pursue the scientific study of orcas, or cetacean behaviour at the very least. I had even accepted an offer to study for a Master's in Marine Biology starting that September.

Dad slid a pint of cider and black my way from across the pub table. It was the evening after a very long day, and we were all sat together inside the Black Bull. The Old Man of Coniston cast his shadow over us, with its dark summit that has no doubt passed its judgement on mortals like me for as long as people have been wandering into the Lake District.

I couldn't touch the pint, and I couldn't stop thinking about everything that had happened either. Seeing the birth of Larry and, later, meeting that sheep farmer and all his collies, and grilling him about sheep farming, lambing, herding and life itself, had been profoundly moving. He had even taken me to his barn and handed me a newborn lamb to bottle-feed. I wasn't three and with my nan any more. I understood what was going on here and why this was so important; that the lamb's mother didn't quite have enough natural reserves of her own milk in her teats, that the lamb needed help from a human hand to survive.

I just couldn't get it out of my head, that feeling of intense belonging that I had felt in that barn and those fields earlier that day. It felt like it was where I was truly meant to be, but what about everything I had worked so hard for up to that point? What about my Master's in Marine Biology? And what on earth were Mum and Dad going to say? Even in my own head, it sounded utterly insane.

As the sun sank somewhere towards the southern edges of Coniston, the whole lake became bathed in an ethereal pastel-pink glow. There was not a breath of wind, so the hills and fells were perfectly reflected in the lake's mirror-calm sheet of water. It was like the whole of the world had flipped itself upside down at the very point where air met water. A sweet metaphor for how I was feeling too, since my whole world was pretty much upside down as well.

Suddenly, I heard an approaching commotion outside. Shouts, whistles, voices and bleating animal noises. Everyone ran outside the pub. A whole flock of sheep were being driven directly through the middle of the village and within seconds they had descended on the pub's garden.

The sheepdogs were soon diving between the legs of the beer drinkers, chasing the scattering sheep under the pub tables and picnic benches, picking off the strays in seconds, and guiding them quickly back into the main flock that was progressing chaotically, but steadily, right through the heart of the Black Bull.

Visually it was an absolute riot of activity and excitement, but I locked in on the incredible calmness of the farmers leading the action from the rear. There was no sense of haste or panic among them, in fact, there were hardly any commands at all. It was as if they already knew their dogs would instinctively solve the problems thrown up by this highly complex series of pub obstacles and that, if they had any hesitation at all, just a word or a whistle from them would immediately arm the dogs with the precise knowledge of what they needed to do. Sheep farming at its most traditional, nothing more than men and collies: timeless and near-telepathic.

I had gone to West Cracroft to further my academic experience, but it had actually taught me more about my own instincts, my resourcefulness and my self-confidence to make sound decisions about my life.

The pub settled back down and the punters all excitedly began reviewing the pictures they had just taken, and missed, of the quintessential Lakeland scene that had swept past us all. I knew what I needed to do. Despite all the hard work to get this far in marine biology, the fact was, the academic study of sea creatures would nearly always be hands-off. All the animal behaviours and life-processes would be outside my arms' reach, both physically in the field, and then again on all the pages I would have written, and the lectures I would have given, if I ever were to make it to the top of the tree in that career path.

It would have been so easy though; especially having taken all the risks and exams, spent the money, and invested the time; to just plough forward with the Master's regardless of how I felt that day. I knew that was probably what my parents would have wanted me to do. They had been desperately worried through that tough patch at school, and seeing me finally commit to something long-term had been a massive source of comfort.

But these are the moments in life where you really have to listen to your gut and find a way to be courageous.

Many people live really happy and contented lives without ever taking risks. Lives where they get genuine satisfaction from making rigid plans that they stick to. I accept that completely, but there are also lots of people who are like me, that have moments like the one I was having in Coniston, and just don't go for it. It's been suggested to me in the past that it was much easier for me to change my life's path at a time when I didn't have the responsibilities of bills, mortgages or kids. That's true, but a counter to that is that the windows of opportunity to truly chase your dreams are only open for fleeting moments, and if you ignore them, put them off, and think you can pick it all up again at some later date, then you run the risk of having a lifetime filled with regret and bitterness.

Of course, I was worried that pursuing farming might have been a disaster, but it was nothing compared with the disappointment I know I would have felt had I not even tried. It wasn't just that I was young, flighty or over-impulsive. I can look back now and still know I was absolutely right to think the way I did, win or lose.

I didn't have a clue why the sheep were being moved or where they were being taken that evening in Coniston, but I knew there was no way I could put all those feelings back in the box that they had just sprung out from. The Master's could wait. I looked up from my pint and told my family I was jacking in everything to become a shepherd.

Chapter Five
Spotty Wellies and Tracksuit Bottoms

'I can't believe this place.'

I had already spent several hours behind the wheel just to get to the farm's nearest village, Betws-y-Coed, a touristy place in North Wales that's nestled on the outer edge of Snowdonia's mountainous crown; but the farm itself was apparently still a fair distance away. I'd found the sign to it easily enough, but half an hour later I was revving my poor little red Mini up and down hills on a seemingly endless muddy track, with no actual farm buildings, farmers, or even signs of human life, anywhere to be seen.

Eventually I skidded into a farmyard and hopped out of my totally impractical car. I had a bright pair of spotty wellies on my feet, bright-blue shell-suit tracksuit bottoms on my legs, and a sparkling new North Face jacket on my top. I guarantee you now, if a stranger had been given unlimited guesses as to what I was doing on that farm, based purely on how I looked that day, 'farm labourer' would've been several hundred jobs after travelling hairdresser, Jehovah's Witness, or weekend-walking-warrior.

I basically looked like I'd just rolled out of the discount bucket at the 'Go Outdoors' Liverpool city branch and headed directly into the hills for my very first day in the countryside – which, to be fair, wasn't massively off the mark. I was 20 years old and starting a career in farming, with no formal experience whatsoever.

In the weeks before my expedition into Wales I had called up virtually every farm in the north of Britain looking for work. I applied for hundreds of jobs I was blatantly unqualified to do and received flat rejections from the ones that bothered to reply. Mostly, though, I was ignored completely. It was hardly surprising given the last job on my CV detailed all my experiences with orcas in Canada. How were they meant to know I was deadly serious about working in British farming? What if this townie woman from the Wirral, who had only had some weird lamby revelation on holiday with her parents, had rocked up and actually caused far more trouble than she was worth? Which, let's face it, wasn't a whole lot anyway; I mean, I wasn't even really expecting to get properly paid. Looking back, it was astonishing I had eventually found someone willing to take me on, even if it was for just one week. I'm not sure I would've taken a punt on me back then.

I still have no idea what the farmer must've thought when he locked eyes on the state of his newest recruit, but I realised pretty quickly why I had got the job. The hangdog-looking Welshman who strode up to greet me could hardly speak a word of English, let alone read my CV.

The farmer was very tall, but quite thin and pale with a shock of jet-black hair. He looked like he could use a bit of love and probably several of my mum's Sunday roasts. 'Alright?' he greeted me gruffly, holding out a hand so thick it was big enough to comfortably swallow mine whole.

'Hiya!' I proffered brightly, smiling warmly. He looked at me like I was an exotic animal freshly escaped from a local zoo.

'Follow me, then,' I think he replied. 'Digs are over 'ere.'

I grabbed my bag and followed him quickly across a long yard and down towards a static caravan that was parked at one end of a field. He creaked the door open, flicked on a gloomy light, and made to leave. 'Codi am pump,' he muttered towards his well-worn wellies, almost as an afterthought.

'Sorry?' I chirped, a little apprehensively.

He stopped, turned to look me squarely in the eyes, a realisation of sorts clearly dawning on his face before he answered in plain English: 'Up and out at five. For some hard fecking work.'

I laid my bag down on the caravan's patchy carpet as he strode back to the farmhouse. 'Well, this is it, Hannah, everything you dreamed of.'

The caravan wasn't too terrible. I'd go on to stay in much worse in those early years of my career, but it was a bloody cold start to May that year. It had one little electric heater tucked into a corner which put out about as much heat as the knackered light bulb in the roof; so, I flicked on the gas oven and chucked some frozen fish fingers under the grill.

Despite what you might think about my naivety, I was for real about this. All of it. Mum was having sleepless nights, though, especially since she heard I was heading off alone to a farm run by a single solo man in the middle of nowhere. I'd had to give her his full address, telephone numbers, local emergency services and GPS co-ordinates for a landing spot for the police 'search and rescue' helicopter; but this was always how it was going to be. I had no knowledge of actual farming practices and no one to ask for help. At some point I was going to have to take a big risky leap into the unknown, and this was it.

Slowly, the place warmed up and my mood brightened. I pressed the fish fingers firmly into a sandwich. Everyone has their first day at work, and everyone has to start somewhere. My somewhere was a farm near Betws-y-Coed (in English: 'The Prayer House in the Wood') and I was about to receive a baptism of fire.

'I'm gonna shoot that fecking dog cont.'

It was just gone five in the morning and the farmer was already on one.

'The dog's called Cont?' I asked, trying to be polite.

'No, the dog's called Tom,' he snapped back, 'and the useless fecking fecker can't see a fecking thing.'

The farmer definitely knew his way around the universal language of swear-words, but was he really going to shoot his dog? I didn't know how to react, so I just sort of laughed nervously, and hoped he was joking.

'Well, what's the fecking point of keeping him? He can't fecking see.'

It felt like a horse had just hoofed me straight in the stomach. I thought of all the animals I had rescued through my life so far. Tom wasn't even an old dog. He had this beautiful shaggy hair, doleful eyes and sweet playful nature; but as he ran off enthusiastically to start the working day, he smacked straight into a closed gate.

I didn't yet know it, but it was a warning of darker things to come.

This was a properly traditional Welsh hill farm. The sheep-farming landscape of the United Kingdom is made up of many varied terrains, but the industry ring-fences them all into three broad sheep-farm categories: hill, upland and lowland. A farm can easily hold a mixture of those environments, but effectively the type of terrain that dominates your acreage determines the number and species of sheep you can effectively keep on your land. In predominantly lowland farms you are going to have lusher, more productive grasses for grazing, and you can choose heavier set sheep breeds that breed faster in the warmer, more sheltered and more abundant fields. Upland farms would include ones around the Lake District and the Pennines, where the land isn't quite as unforgiving as the high fells and hill farms, but you are still going to need a hardier breed, which you could crossbreed with lowland sheep varieties if you have more comforting pastures for them to feed on through the cooler months. On hill farms you are generally looking at the harshest, most hardcore environments: windswept open fell where the soil is relatively poor, even acidic. Only the hardiest breeds can survive out on the hilltops of Snowdonia in Wales, and the Highlands and Islands of Scotland, but you are still going to need access to lower-lying areas of shelter to bring your sheep down when it starts to get really cold. Even the hardiest sheep in the land will struggle or die if left to fend for themselves in sub-zero conditions and thick snow.

This hill farm wasn't quite as rugged and raw as others found at higher altitudes elsewhere in the Snowdonia National Park. He had a bit of moorland grazing on his highest patches of land, but plenty of more comfortable pasture too. It was a lot closer to what we have in the Cumbrian uplands, and a lot friendlier to farm as a result, but that was where the friendliness ended.

The farmer would scream at me in Welsh all day long, and then scream at me again for not understanding a word he had just said. Now, I obviously have no problem whatsoever with the language of the farm. It was his farm and I was in his country after all, so he was well within his rights to speak Welsh. Also, when he did switch to English, just to remind me of fundamental facts such as: 'You haven't got the first fecking clue what you're fecking doing,' I could sort of take his point: I didn't have a 'fecking' clue what I was doing, and I imagine, when you're trying to get a job done quickly, and you've got some complete muppet continually making the most basic mistakes and incessantly asking questions about the most trivial-seeming things, it would quickly become very annoying indeed. However, I had hardly duped him into thinking I was going to be anything but a rank beginner. It wasn't my fault he hadn't read my CV and, given he steadfastly refused to explain what I should actually be doing anyway, or why we were doing something a certain way, it wasn't like I had much chance to learn by anything other than trial and error; and error, and error again, and yet another error.

As the days dragged on I couldn't help but feel there were times he was deliberately setting me up to fail, almost like he was taking some sadistic pleasure in seeing me struggle. One day, with the rain absolutely pouring down, I can remember being quietly pleased with myself for having had the forethought to buy some brand-new waterproof trousers. Diligently, I tucked them inside my wellies at dawn, with the farmer watching on silently. All day that day, my feet were absolutely soaked through and freezing cold, as water ran down my trousers in hundreds of tiny rivulets and sank directly into my boots. When we finally finished work, I pulled my wellies off, poured out about a pint of rainwater onto the shed floor and turned to hear him remark: 'You shouldn't have tucked them trousers into your boots. That's why you're so wet,' with a hint of a smirk.

'Well, why didn't you tell me that when I was bloody doing it?' I fired right back.

'Because I want you to learn the hard way,' he replied bluntly. What a dickhead.

That first morning on the farm we cleaned out the animals together before heading into the fields to move the sheep and lambs. He started to gather them up, using me as an extra dog to squeeze them all down into the holding pens beneath us. Once they were down in the pens we 'shedded' off the lambs, separating them from their mums by way of a funnel called a race.

I say lambs. They were already about a month old, and one thing you realise pretty quickly is that sheep put on mass at an extremely fast pace, right from birth. They might be only three or four kilos when they are born, all cute and cuddly, but after just a few weeks they can be as much as five times that size, depending on the breed and the farm, and almost resembling fully-grown adult sheep.

These guys were absolute beasts, so shoving them around from pen to pen was a deeply physical task, especially for someone who had never done it before. Once we had all the super-size lambs penned inside a yard with a concrete floor and breeze-block walls, the farmer shouted over: 'Right, catch them lambs now and bring them to this table.' He hammered his fist down on a big wooden block.

I grabbed the first one, gripped its fleece hard and heaved it over to him. 'Well, turn it, then,' he grumbled, in a way I had come to accept meant I should have been able to understand his every wish telepathically.

Struggling with the weight, I flipped the lamb onto my knees and held it across the table, presenting its underside. The farmer immediately reached down, produced a devastatingly sharp knife, and sliced its tail clean off.

I gasped. The lamb cried out in pain, and booted me in the leg so hard the bruising would've felt absolutely explosive, were I not already completely numb with shock.

'Oi, grab the lamb and mark it up, then. Come on, get on with it,' he snapped.

I looked to the lamb, its tail now just a bloody stump, and marked its side using a stencil dipped into a tin of marking paint. It was now officially a part of this man's flock.

'Over the wall then, now, get going,' he intoned gruffly.

I picked up the freshly sliced lamb, heaved it up and over a small barrier, and down into an empty pen. I looked around wide-eyed. There were dozens and dozens of lambs to get through.

'What, but why . . .?' I tried to stammer out, before the farmer cut me short: 'Look, this needs to be done, alright? It's welfare. We haven't got time for fecking about talking. Grab the next fecking sheep.' He pointed out a lamb with the sharpened point of his knife.

Every day was the same from that point forward: gather the lambs and ewes in the morning, split them up, then spend the afternoon running around grabbing jumbo lambs to get their tails chopped off. No pain-relief. Just off with the knife, over the wall, and right onto the next.

That day was a huge shock to the system, but I was viewing it as someone whose only prior physical contact with a lamb had either been highly sanitised, at the petting farm with Nan, or completely one-dimensional, seeing Larry getting born happily into a field and scamper off into a barn. I knew nothing about the realities of the sheep-farming process, or the sometimes brutal nature of life on the farm.

Unbeknown to me at the time, the farmer wasn't just some madman mutilating his sheep for absolutely no reason. He was 'docking' the lambs' tails, the name given for a fairly routine procedure carried out early in a sheep's life, where the lower section of a lamb's tail is removed to prevent it from becoming coated in their own faeces. The poo attracts swarms of flies to an undocked tail, then their maggots and, ultimately, a deeply unpleasant affliction called 'fly-strike', where the sheep are, quite literally, eaten alive.

Sheep were one of the first wild species of animal to be successfully domesticated by humans. We believe a shaggy-looking upland beast called the Mouflon was the original breed to be effectively farmed, right back in 10,000 BC in an ancient place called Mesopotamia, found deep within the 'Fertile Crescent' of Western Asia. Since then, successful selective breeding strategies have created a vast spread of sheep species that are near unrecognisable from that original Mouflon. The different characteristics of all the many varieties today have allowed farmers to create animals with features tailored specifically for their land, environment and the domestic market, with the ultimate goal of developing the thickest cuts of meat for eating, and the densest fleeces for our clothes. None of that necessarily builds the strongest animal for the natural world though, which is why we must take so much care of our sheep throughout the seasons, to make sure they stay fit and healthy. There are some who believe farm animals could be freed directly into the wild if we all gave up eating meat tomorrow, but the truth is our farmed sheep could never survive without us, especially when something as simple as a humble fly can get them into life-threatening trouble.

Once the poo is in the tail it really doesn't take long for the flies to find it. One female bluebottle can lay 250 eggs alone, and within 12 hours that poor sheep will be completely infested in maggots. The maggots are absolutely awful. The first indication is a discoloration in the wool, a dark-green stain appearing around the bum and chest. That is when your heart absolutely sinks to your wellies. Pretty soon you'll actually notice the maggots have started to separate the fleece from the skin of the sheep and if you don't get on top of it then, they'll soon begin to burrow deep into the animal's flesh. You feel so awful for the sheep. All you can do is treat the area, which means first removing the fleece as far as the maggots have extended, sometimes right up the sheep's back and towards its neckline. Then you must treat their red-raw skin, and any wounds, with antiseptic; I always administer additional antibiotics and a painkiller on top. After that, it's just a waiting game to see whether any areas of that sheep develop bacterial infections; if necessary, they may even need to be put out of their misery and humanely dispatched.

Or, you can just 'dock' their tails. Give them a morning's discomfort when they are young and dramatically reduce the chances of this fly-strike from ever happening at all.

The laws over the methods you can use are a little vague. In Wales and England, the 'Permitted Procedures Regulations' state that if you are not going to use a rubber ring – the most common method, which involves applying a constricting ring of latex, between two and seven days after birth, to choke the blood-flow and allow the tail to drop off – then 'another method' can be used later 'if an anaesthetic is administered'. Within the industry the non-legal advisories are much clearer: the Sheep Veterinary Society says that using knives causes considerably more pain than the use of the rubber ring and recommends that the use of a knife to dock 'by anyone other than a veterinary surgeon' should be 'prohibited'. If it isn't explicitly a legal choice on whether or not you should use a knife, then it certainly feels like a moral one and, regardless, the use of an anaesthetic was, and is, mandatory. I certainly never encountered that method again.

If there isn't such a high risk of fly-strike then farmers won't actually dock their sheep's tails, and the hardier breeds that are going into the gnarlier hills will often keep theirs too, as the strong winds up high can actually give the older ewes mastitis – a nasty infection in their udders that causes intense pain and also takes antibiotics to treat. Even when they do dock, good farmers will still make sure they leave enough of a tail to cover the anus or the vulva, just to give them the protection they need from the elements wherever they are sent to graze. Also, as the price of wool continues to plummet, a new breed of sheep called the 'EasyCare' is coming into the market, which may yet eradicate the need to dock at all. The fleece of the EasyCare sheds naturally, regularly and completely, right into the field. If that breed takes hold in Britain, then docking could well become a thing of the past anyway but, for now, knives are best avoided, and if you are unfortunate enough to witness fly-strike on a sheep, then make sure you don't have white rice for dinner.

Obviously, though, I knew absolutely none of this on my first day on that farm in Wales.

*

Never mind the verbal abuse, a lot of people would've walked out after a day like that – I was only working for food and board, after all – but I reminded myself that this was only the first step in a very long journey and, just like all the old school rules, I didn't have to agree with everything (or ever use it on my own farm, in the very distant future), but I did have to hang in there for now, just to try and gain something from this experience.

Looking back, I think the hardest thing was that, as instinctively uneasy as this man and his practices were making me feel, I didn't actually know if they were really wrong or not. This was my first experience of farming, after all; I had no point of reference and was effectively blinded by my own ignorance.

I tried to re-program my brain. This was a real-life, fully-functioning commercial farm that needed to make money to survive, and if I was going to survive too, I would have to try and hit pause on all my judgements, and park any expectations I had. I resolved to just take each day as it came and try to stay positive. I would do as I was told – this man was not going to break my spirit.

There was one lesson that I did need to learn 'the hard way'. The shepherd's eye is nothing like the eye of the tourist. I remember driving towards that farm before the job had even started, passing the mighty peaks of Snowdonia, and feeling really excited. It felt amazing to be working in a place that was so visually spectacular; but as much as those gullies, waterfalls, cliffs and rocky outcrops would've made amazing photographs for our family's holiday albums, you only have to fish the heavy corpse of a sheep out of them once to realise they are all death traps.

Nature is remorseless and unforgiving. Fly-strike aside, new-born lambs and pregnant ewes will also regularly have their eyes and tongues pecked clean out and eaten by magpies. Rogue foxes will predate on lambs and vulnerable sheep. Even moles burrowing under your fields can cause the earth to collapse when you are moving heavy farmyard machinery, and will potentially break the ankles of your animals. The molehill earth itself, if bundled up in hay bales and fed to your sheep as sileage over winter, can also give them a terrible bacterial disease called listeria, causing their brain to swell and killing them very quickly if left undiagnosed.

Magpies, foxes, moles . . . all of these are quite cute, warmly regarded creatures outside of farm life, animals I would have been thrilled to encounter at any point before coming to the hill farm, but they are all sheep killers in their own way.

Those challenges from nature need to be controlled on your farm – where and when you can, and within the law – but the reality of farming life in Britain is that there are an awful lot of people squeezed onto quite a small island with comparatively few places where sheep farming can happen successfully. Often, we will be working cheek-by-jowl with wild creatures that will harm our flock if they are allowed to act with impunity, and many of the natural controls of these animals, apex predators and the like, disappeared from our landscape over a century ago. In my opinion, successful farming and successful natural landscapes can co-exist, and ideally work in harmony, but there are times when sheep farmers simply have to tip the balance towards their vulnerable flocks. The profit margins in farming are slim, and getting slimmer, and one bad winter, one bad fly-strike or one bad predatory attack can make the difference between being a profitable farm or being doomed to failure.

Farmers are not above the moral code of life, though.

You can't just go around killing everything that's giving you a problem, whether it's the wildlife, your dogs, or even your sheep. Good farmers have both compassion and environmental awareness. If your sheep have just become numbers, and your dogs just another piece of soulless farm machinery in the farming process, then it is probably time to consider what you are doing in the job. When you are on a farm where the farmer has lost that elemental sympathy, you can be sure they are also well on their way to losing their passion for the job too.

I don't actually think the farmer on that hill farm in Wales was a bad farmer. His flock was a healthy one and he clearly cared for them greatly. I think part of it was that he was just really set in his ways, and his ways were very old-school and pretty unchanged through the generations of men that had farmed his land before him. His docking methods aside, I remember him also once saying he'd turn the wireless on for me during a lunch break, but when I reached for my smartphone (assuming he meant the internet), he just flicked on the radio (tuned to a Welsh-language station, naturally) and walked out the shed. I hadn't heard someone say 'wireless', when they meant 'radio', since Great-grandad Bill.

I think he was probably quite unhappy and lonely too, but I still don't think there is any excuse for threatening to shoot one of your dogs. As much as I had to put my preconceptions and judgements to one side, I still tried to get that dog, Tom, rehomed. My mum found a charity for collies that no longer have the ability to work, and I asked the farmer if I could take him but, ultimately, he wouldn't let him go. 'You are not taking my dog off my farm,' he said, in a tone that made it very clear there was to be no compromise. There wasn't really much more

I could do, and I can't say one way or the other whether Tom ended his days at the barrel of that man's gun or not.

Today, I always aim to give the animals in my care every chance of survival. You would be surprised at just how close to death they can seem to be, and yet still be brought back to life with a bit of loving care and attention.

Years after that first job, on a friend's farm, I helped birth a sheep from one of his very best ewes. When it arrived, it was listless with only the very faintest of heartbeats. It certainly looked like it was going to die, and although my friend was gutted to lose a lamb from one of his finest sheep, he was pretty direct about the next course of action: 'Hannah, that one is going to die, best we move on.' I knew it had a chance, though. It is something I feel instinctively just by the look of an animal, that there is enough going on deep down in there to just give it some hope of survival. So, despite all the vital signs draining in front of my eyes, I quickly began massaging its heart and organs, right there, on my knees, in the field.

'Hannah, just give it up,' my friend shouted some minutes later, when there hadn't been any obvious signs of progress, but I persisted, swinging it up and down and even dunking its head in the water trough, to try and shock its heart into life. Slowly, I could feel it recovering; later that evening, I took it back to the farm sheds, put it under a heat lamp and gently tube-fed it by hand. That lamb lived to be big and strong, and, I'm proud to say, it became known as 'Hannah the Lamb'; a top-class prized ewe back in my friend's breeding flock.

I'm not completely soft, though. Some animals really won't make it, and you have to use your own judgement and experience to figure out which ones are worth investing a concentrated effort into. For me, as a farmer, I need to know I have done absolutely everything right, but the facts of life are that some animals are simply destined to die before their time; often, the most ethical thing is to euthanise them yourself. When that happens, you don't make some big, boastful public display of your bravado, you take it away somewhere quiet and you ensure that the death is as clean, dignified and peaceful as it possibly can be. Farmers have the right to take life on their farm for welfare reasons but, for me, I prefer to call the 'dead man', the person licensed to first administer a stun, and then a swift bolt to the head. Again, all direct, ethically sound, and with the minimum amount of suffering to the animal.

Within days on the Welsh hill farm I could feel my mindset changing as I stopped being the child who loved all creatures great and small, and would do absolutely anything to keep even the most lost of lost causes alive. I slowly began to understand that there are times you have to put the flock, and the farm, above all else. The biggest fact I would have to learn in farming was that we are ultimately in the business of raising animals for food, that death will come regardless, and that is how I get the money to put food on my own table.

I eventually finished my week on the hill farm with bruises so dense and black that it looked like I'd been infected with the bubonic plague down my whole left side.

I would never ever do anything like that again, but I'm strangely glad it happened the way it did. I probably did need to learn a few lessons 'the hard way', and that first job, that confusing fiery crucible, hardened me up from the very start. I knew I was ready for anything any farm was going to throw at me, and I wasn't quite so ignorant any more. Right there, on my CV, I had proper farming experience.

I gripped a stiff yard-brush with my bloodied hands and swept a pile of severed lambs' tails into an empty sack of sheep feed. '*Diolch*, Hannah.' Thank you, the farmer said, holding out his hand for me to shake. I wasn't completely awful after all. Three weeks later, he even invited me back.

Chapter Six
Speaking Sheepdog

From the moment I had seen those sheep farmers in Coniston, using their dogs to move sheep up the lanes and down through the pub, I had become completely obsessed with the idea of learning to do it all for myself.

I wasn't really given any opportunity to learn in Wales. I was just one more dog in that farmer's pack, albeit one that didn't understand any of the calls or whistles, or any of the theory, or really that much about sheep either.

What I had gathered, though, was that the whistle was very important, and that I needed to learn. So, given my sole sheep-farming contact had about as much interest in teaching me to whistle as he did in even speaking to me at all, I turned to that infinite educational resource: the internet.

A week after ordering, a brand-new shepherd's whistle dropped into my mailbox. I ripped open the package and held the little metal instrument in my hand. It looked like no whistle I had ever seen before. It was like half a UFO. A small flattened semicircle with a slot along one edge and a tiny hole in the top. I placed it straight into my mouth and blew. Absolutely nothing. Just hot air passing over steel.

There were instructional videos everywhere on YouTube detailing the correct teeth and tongue placement, and the way you should blow to make a noise; and all of them ended with an ear-splitting shrill that could well have been heard in outer space.

I resolved to tie that whistle round my neck and try to get it to work in my every waking hour. Answering emails? Try the whistle. Cooking dinner? Try the whistle? Reading? Whistle. Driving? Whistle. Thinking? Whistle. Toilet? Whistle.

Getting any noise out of that whistle became my sole focus for days on end, and soon it became emblematic of my entire struggle in farming. How could I ever contemplate becoming a professional shepherd if I couldn't even blow the bloody whistle? Any advance on silent wind would represent a giant leap forward for me in proving I properly belonged on my new life's path.

'Peeeeeep.'

I will never forget the moment that piercing note of pure beauty tore into the air like a raised fist. Okay, so it might've sounded little louder than a lamb's fart, but it definitely was an actual change of pitch from just inaudible moving air.

Don't move that tongue a single millimetre now, Hann, I thought to myself, freezing my whole head, and blowing once again. The peep became a tweet, the tweet became a whistle, the whistle became a roar. I'd cracked it, and I proudly coupled the whistle with a fancy new leather lanyard and tied it around my neck like an Olympian receiving their podium gold; freshly minted and ready to do everyone's heads in by blowing my sheepdog whistle all day every day for the next few weeks.

That was it. I really was ready for the next step.

I now needed to find someone who would actually take the time to teach me how to manage sheepdogs properly. A bit of research, a few calls, and it quickly became clear that all roads were leading to one Lake District-based Scotsman.

*

Derek Scrimgeour is a sheepdog-training legend of the highest order. It is hard to overstate his achievements in the world of training, handling and breeding; his dogs have quite literally been sold right around the planet. He's taught hundreds of people to successfully shepherd, written a seminal sheepdog-training book, appeared in television shows worldwide, produced training DVDs in English, French and German, captained the England squad in the World Sheepdog Trial Championships, and is a former English, Irish, and Finnish Sheepdog Trial title-holder in his own right; and yet he is still one of the most modest and understated people I have ever met.

I couldn't have been more ready to meet him when I eventually did, but by the time I had hauled my Mini through yet another white-knuckle journey to his picturesque farm in the Lake District, I was a little taken aback when he immediately asked, 'So, why have you got that whistle wrapped round your neck then, Hannah?'

I looked down at my shiny shepherd's whistle, sat proud at the end of its twisted leather lanyard, but before I could think of something clever to say, Derek – I think helpfully, in his own mind at least – dropped this thunderbolt: 'Don't worry, you're not ever going to need that here.'

Are you actually fucking kidding me? 'Oh, yeah, I err . . . ha, ha, ha.' I tried to laugh in a way that I hoped would demonstrate I naturally knew that all along, and that I had hardly even noticed I was wearing a shepherd's whistle, but deep inside, I absolutely boiled with embarrassment.

As soon as his back was turned, I flicked the offending whistle directly into the neckline of my top, hiding it from his sight completely, but not from my mind. He turned back and I smiled at him; between ever-so-slightly gritted teeth.

Derek Scrimgeour looks just like a country man; a wizened Scot with gently weathered features hewn from a lifetime spent facing into the weather. His reddish wind-whipped cheeks and nose meet kind and comforting eyes. Behind his giant reputation, Derek holds the most unassuming, warm and gentle nature, but he did not waste his words or suffer fools. When he spoke, you knew to shut up and listen; and my goodness, what a man he was to just sit and listen to.

When Derek was on a roll you could spend many hours quiet in his company, just soaking up all the wisdom that fell from the mouth of this sheepdog-handling sage. Derek could speak so well about so many aspects of farming, explaining everything in simple terms that were both easy to understand and still detailed, but he would always finish with a lesson that was well worth learning. It was night-and-day compared with my experience on the Welsh hill farm in the weeks before, and my upcurve in understanding was set to be meteoric.

I don't ever recall Derek raising his voice. He just exuded his authority naturally, and was one of those people that you just desperately did not want to disappoint. It was a calming presence he took with him into the fields. You could be with him out there in the complete bedlam of a grand gather, with hundreds of sheep moving down the fells, herded by a dozen excitable dogs, and he would never flap, or lose control, or shout. He always seemed in total control.

Perhaps his farming manner shouldn't have come as a huge surprise given he has been trialling dogs in high-pressure international competitions for over a quarter of a century but, as much as I am sure there are some farmers out there who might be able to compete with Derek in terms of his experience and knowledge, I've still not met anyone who shares that elemental composure. Derek is just Zen.

Derek initially invited me up to his farm for a couple of weeks' apprenticeship, but I ended up staying and working for him for nine months. My food and board were covered, although I did have to share my tiny caravan with a crazed peacock that would leap onto the roof every day before dawn, waking up the entire fleet of almost thirty sheepdogs with its incessant crowing. Thanks to that peacock, down on Derek's farm I was actually waking up even earlier than I had in Wales.

I had brought along my own collie dog, Dan, who had shown an early interest in sheep and seemed very much at home in the farm environment (secretly, I held quite high hopes that we would one day work together in a farming partnership). I was also really pleased to discover that there was other human company. Aside from his wonderful wife Helen, I was also working and training really closely with two other women of a similar age to me: Derek's daughter Rachel, and a Finnish woman called Annie. Both were just as keen as me; Rachel was obviously naturally gifted, being Derek's daughter and all, but Annie was too, and would go on to successfully breed quality sheepdogs.

Derek would happily admit that women were often better at working sheepdogs than men. He argued they were naturally more sensitive, encouraging and sympathetic to the dog's innate needs, whereas men can have a tendency to just steamroll the dog into submission under a series of noisy demands.

That's not to say that men can't be sensitive too, or that women can't turn the heat up when necessary, but Derek was working off evidence gathered from training shepherds over several decades, and that was just what he had generally noticed; alongside the fact that if a particular dog doesn't like being shouted at, then that'll usually be the last bit of work you'll ever get it to do.

I know Derek would probably disregard this, but it all showed just how progressive he was as a farmer. He was happy to work with young people, or women, or outsiders; and by 'outsiders' I don't just mean people from other countries. Derek firmly believed that shepherding could be taught to people from places well outside of farming. In short, he was willing to believe in me. The debt I owe him is simply enormous.

Derek would give you lots of little jobs to do with the dogs, slowly building your confidence by gradually increasing the difficulty and giving you space to work alone, but he was always there whenever you needed his help, as if he had actually been keeping one eye on you the whole time.

Day one, lesson one, was a bit of an understanding of Derek's main mantra, which reads: 'A working sheepdog needs three vital traits: temperament, ability and stamina.' All three I felt Dan might potentially have had, but it soon became clear that it just wasn't going to work out as I had hoped.

Dan was five, still young in domestic dog years, but actually quite ancient in terms of his ability to learn to herd sheep from a standing start. 'You can't teach an old dog new tricks,' the tired cliché runs, but sadly it was going to be true of this particular sheepdog.

Everything was just one big game to Dan. He would very happily run in circles round and round the sheep all day long.

It was the entry-level exercise for all young sheepdogs, but he never once showed any sign or interest in progressing from that very basic point. He just couldn't seem to get into a working frame of mind and, sadly, he had no functional value as a sheepdog as a result.

I was sat at Derek's kitchen table after my first week, clutching a steaming cup of tea between my palms. 'I just don't understand what he's not getting, Derek, I feel like I'm doing everything right.' I looked up to him for reassurance. 'I mean, I know I am, because when I train with your other dogs we move the sheep in the right sort of directions, and seem to make progress, but Dan just . . .' I began to trail off.

Derek leaned back in his chair, his face the picture of reassuring kindliness.

'Do you want me to be completely honest with you?'

He had this habit of framing bad news as a question first, instead of just blurting it right out.

'I think he's probably already found his peak, Hannah, and if you want to continue with Dan, you are going to waste a lot of time with a dog that is ultimately never going to do what you need him to do.'

It was a tough one to take, but Derek could hardly have put it any clearer. I sipped my tea quietly. I knew then I had probably put a bit too much faith in Dan's abilities, but I also knew just how incredibly expensive top sheepdogs were. Even the part-trained young pups on Derek's farm were worth considerably more than my car, and if I could somehow scrape together the money for one of those I would still be looking at several years' training before I could even start working, and earning, with them.

It did feel like a setback, but on the other hand, Derek was right: there was absolutely no point persevering with Dan, and it was always going to be better to know that right from the beginning. At least I was going to save my time.

Dan did stay around for a while, just as company, but even that underlined just how much this really wasn't for him. Dan was fully satisfied with just hanging out at the farm, whereas what I really needed was a dog with a real hunger for the job; but finding *that* dog now felt very far away indeed.

There often seems to be this undertone in society that jobs in farming are reserved for thick people. That 'Dr Jackson, whale scientist', has a hell of a lot more kudos than 'Hannah Jackson, farmer', and I just don't get it at all. I think it makes far better logic to presume that someone who is producing the food that keeps you fed and well is, at the very least, of equivalent importance to a person who monitors an animal species that you'll probably never even see. Besides, I actually found learning how to handle sheepdogs, keep sheep and, ultimately, run a farm, far more demanding and multifaceted than I ever did studying animal behaviour academically. Society places a warped value on things all the time, though; we are as seduced by the letters before and after people's names as we are by pseudo-celebrities who are famous for not really doing anything much at all. It's a reason why it is always better to set your goals and your barometers for success for yourself, and try your best to just ignore all the other noise.

I actually had a slight leg-up at Derek's thanks to my Animal Behaviour degree, and the sheer amount of time I had already ploughed into trying to understand animals, especially dogs. When it came to learning how to actually work a sheepdog, though, I was pretty much starting from scratch.

Luckily, Derek had patience in spades, and my pathway had been well trodden by the many hundred students that had come before me.

Fortunately, my time with him spanned across an entire summer, so we had long full days of training and work. The sheepdog 'course', if you could even call it that, and the actual work of training the dogs themselves, were completely intertwined. I was there to work *and* learn. Derek farmed a hill farm as well as training dogs, so I was also able to get a far greater knowledge of sheep farming with him, and take the practical skills of the training straight out into the field. It could scarcely have worked out better.

I started with the basics: learning how to train the very youngest of the collie dogs, starting at just five months old, in the most basic of moves. Surprisingly, given what you might think from watching people work dogs on television or in the fields, you actually start them off without any calls at all. I was taught by Derek that dogs actually predominantly learn through your body language and positioning, especially when they are young.

There are not many collie pups that won't instinctively want to charge straight at the sheep the first time you work with them, so you are going to need a shepherd's stick, or just something long and visible. At Derek's we often just used this big bright-blue piece of pipe, but whatever you use, it acts as an extension of your arm, and can be deployed to push a dog back out of the sheep flock, if it does go bolting forward into them. The far shrewder, and ultimately more useful, trick is to stand in the parts of the ground where you don't want your dog to be and use your stick, held out wide with your arms spread, to occupy as much of that space as possible. Derek calls this, 'putting pressure on the ground', and learning this method and its reasoning was as central a part of the dog's training as it was my own.

Imagine the space within a training 'ring' acting almost like a moving clock face with the sheep positioned right in the middle. If you were hovering at '12', and the dog was directly opposite you at '6', the dog can go one of three ways: clockwise round the sheep, counter-clockwise round the sheep, or straight forward towards the sheep. If, in this example, I wanted the dog to move clockwise, towards '9', I would hold out the stick and walk towards '3', placing 'pressure on the ground' in that area with my presence, and opening up a more natural pathway for the dog towards that '9' spot. When it all works well, I should be able to move the dog around the opposite side of the clock face to myself, with the sheep on the move, but always occupying that central ground between us. The exercise is fundamentally all about the dog learning how to maintain a working distance, while continuing to move sheep with those lightly pressurised circular movements around the perimeter of the clock face.

It hammered home one of the big behavioural differences between sheepdogs, like collies, and all the other domestic dog breeds. In a home environment your dog is probably going to be most comfortable when it is closest to you, finding its happy place at your heel, but when you are working sheep, you need your sheepdog to be entirely comfortable operating away from you. The sheepdog should naturally want to work those sheep, so all you are doing is gently teaching them the most effective way to build on their own biological instincts, while reminding them that respecting first your movements, later your voice, but ultimately your absolute authority, is fundamental to having an effective working relationship.

One of the hardest things to teach the dogs is that there isn't a definitive ending to that first training exercise. You're not moving the sheep into a pen, shutting the gate behind them, and standing down the dog; the exercise ends when you call it and, more often than not, the young dogs will not want to finish.

I can remember in those first few weeks having a real problem training this dog that just did not want to stop circling the sheep. I kept shouting at it and whacking the blue pipe on the floor, but it would not listen at all. Eventually I managed to grab hold of the disobedient dog, stick it on its lead, and gruffly frog-march it back down to its kennel.

I walked back up to the gate to re-enter the training ring to carry on with a new dog, but Derek was waiting. 'Go back and get that dog,' he said, calmly.

'Why?' I responded, with obvious exasperation in my voice. 'It just took me ages to catch it and get it back.'

Derek leaned in on the gate. 'Because, Hannah, that dog will never finish when you ask it to, if you finish its exercises on bad notes like that. You've got to bring it back here now and let it have another go.'

I went and retrieved the dog from its kennel and walked it back up. Derek was actually making a very valuable point: if, every time I work with that dog, we have this massive battle to get out of the ring, it is just going to think, *Well, what's the point in ever stopping for her when I'm asked to? She's just going to shout at me anyway*. Pretty soon you'll forever imprint the idea in that dog's head that finishing any job is always a negative action, and then it quickly becomes useless as a sheepdog.

I restarted the exercise. Taking up the positions on the clock face, allowing it to recircle the sheep, until I called the finish. It didn't come back right away, but it didn't take as long as it had before, and when the dog did eventually come over I gave it lots of positive reassurance – hugs, pats and 'good boys' – reinforcing that good piece of behaviour in his mind, so he'll hopefully remember that a big love-bomb is coming when the work is called off.

Out of the training ring I graduated to working with the juvenile dogs – usually eight months and above – that were just starting to work the sheep in bigger fields. Once you've taught the dog to circle with control, and change direction under pressure, that's when you can start throwing in the most basic of the calls: 'come by', which means you want them to go left around the sheep; 'away' to get them to go to the right; 'walk up', which means walk up behind the sheep; 'that'll do' to mark the end of a move, or a job; and 'lie down', which is your speed control, to get the dog to pause and create some space between itself and the sheep, and sometimes also the 'panic button', for when it all gets a bit chaotic and you need some time to regain your thoughts and decide where to go next in the grand game of sheep chess. 'Lie down' also gives the dog a momentary reminder that you are there, and that *you* are the one in control, just as it looks like it might be on the verge of going rogue and leaping in on the flock.

If a dog takes to the calls well, you can start to phase out some of your movements and the need for that constant physical pressure on the ground; but, even with the best dogs, there is rarely a time that you aren't constantly on the move during sheep work. Those very basic moves, Derek's 'ground pressure' learned with the very youngest dogs, will forever be an important weapon in my arsenal as a shepherd.

Derek always underlined that the 'tone' you use is as important as the spoken words themselves. For example, for your basic moves, your 'come by' and your 'away', you almost want to sing the instruction in a welcoming way. You are encouraging the dog with your tone: 'I welcome you to come left' or 'I welcome you to come right'; and the same goes for 'walk up'. But 'lie down' should always come out that little bit more assertive, sometimes even with an aggressive tone in your voice. Especially if they don't heed that call right away.

There are other functional calls you just learn to throw in at certain situations. If, for example, I want Fraser to isolate a sheep, in our terms 'to catch it', say if it was lame, or heavily pregnant, then I'll lead with an urgent and repetitive drill of: 'Get it! Get it! Get it!' Or, if the dog has passed over a sheep without realising, I'll get them to first lie down, and then I'll tell them to 'look back', before they get up and go looking for that individual. Another I tend to bring out if the dog is a bit overexcited, a little too eager and full of beans, is just a calming: 'Steady, steady, steady,' in quite a low humming tone, intended just to get them to drop it down a gear and take a beat.

When I'm way out there in the wide-open fell, I'll also use a lot of encouraging 'ssssh' noises when my dog is close, but at far range, I'll use various 'whistles' (thank God, there eventually were some bloody whistles, although I did end up just using my fingers, and that steel whistle and lanyard got permanently retired to a drawer somewhere) as, once mastered, a good, powerful and varied whistle can communicate many of the same commands as the verbal ones – and, critically, a whistle will travel over far greater distances than a shouted call ever could. That is essential when your dog is working up to half a mile away from you, or even if it is somewhere out of sight altogether.

You realise pretty quickly then that you can't just be handed a sheet of paper with the main calls and expect it all to be alright. Also, you soon find out that you can learn as much as you like about what *you* should do and say as a shepherd, but if you haven't got the right dog for yourself or if, like Dan, the dog is already past it, then it doesn't matter if you are Hannah Jackson or even Derek Scrimgeour, you are never going to be successful with that dog.

Dogs, like people, have radically different characters from one to the next. For example, some dogs want to be going in at the flock of sheep all the time: their wolf instincts dominate over what you actually need the dog to do to successfully complete the job. If a dog leaps into the middle of the sheep, he might well isolate one individual, as he would do on a successful hunt, say, but the rest of that flock will scatter to all parts and the job of herding the sheep back together has to start right again from scratch. You can help dogs that have that tendency by penning the sheep behind a fence, creating a physical barrier that the dog just cannot pass through, while you try and train that animal to maintain its distance, but ultimately, that dog is going to need a master who is willing to assert themselves over the will of the dog, and that can be really hard if it doesn't just come to you naturally.

At the other end of the scale you might get a dog that is actually quite nervous: a bit shy and standoffish from the sheep. You might then begin its training in a smaller area, maybe half the size of a conventional ring. They are naturally a little closer to the sheep then, but still have some room to back off if they really aren't feeling it. Slowly, they should learn to assert their will without the sheep being able to scatter away too far; but again, if you go into that situation all guns blazing and shouting at the dog, it will shut down completely. That's when you need to be a calm person yourself, to suit the needs of that dog.

You might be able to adjust the habits of a dog through time and training, but ultimately the actual character and personality of a dog is written in stone from birth, just as it is with their human masters. Certain habits will always be there, between both dogs and people.

It is another reason why it's so important that you spend time getting to know the dogs you are working with – so you can tailor your approach to their specific needs – but it also pays to realise that there are some dogs that, no matter what you do, just aren't right for you. I have had dogs that I just couldn't work with. I remember one in particular that was just incapable of being calm at any point, almost psychotic in its behaviour in the field at times, and so hard-headed that I just felt I couldn't get through to it. In the end I sent her off to a dog trainer in Yorkshire, someone with a different temperament to me, and they hit it off right away. That dog was never going to be world-class, but there was clearly more potential in there, waiting to be released, with someone else. I know people that have actually ruined good working dogs by refusing to accept the fundamental fact that you can't change the core of their character, or yours, through sheer willpower alone.

One thing is for sure: all dogs have to know that ultimately *you* are their master, and whatever route you feel you need to choose to press that point home, the outcome *must* be the same. You are the boss, and they absolutely have to listen to you.

Pretty much from the start of my time with Derek, I heard him and the other shepherds talking about the 'eye' of the dog. For months I didn't have the first idea what they were talking about when they'd look at a sheepdog and say 'good eye' or 'that's too much eye', but what I eventually discovered was that it was this almost trance-like state that falls over certain sheep-dogs whenever they see sheep.

It can be quite subtle – just a bit of 'eye', where the dog seems to momentarily lose itself while staring at a flock – but sometimes it is deeply intense, where their 'eye' is absolutely fully locked on the sheep and there is nothing else in their world other than those sheep in front of their face. That's when a dog has 'too much eye'; when you just can't pierce through their sheepy stupor and get your voice, your authority, back into their head. In extreme circumstances they can sometimes become so singularly focused on staring at sheep that they won't actually do any work for you at all.

The very best dogs do have a powerful 'eye', though, and the best shepherds are the ones that can dangle them from that thread, teasing their instructions into their conscience, while allowing the dog to assert itself on the sheep. I've seen some dogs move sheep just with the intimidatory power of that look, their 'good eye' alone, but you'd better hope those dogs can still hear the voice of their god though, otherwise whatever job you are trying to do will soon descend into total carnage.

Probably my biggest disaster at Derek's was the day Annie and I were bringing over a hundred sheep down from the fell beside his farm. The plan was to fetch them to a junction in the farm track, with one route leading back up over a small beck and onto the farm, and the other straight down the road in the complete opposite direction. Annie had a young dog called Alien. Her dogs always had absolutely bizarre names. I eventually took a dog off her myself called Butch, probably the only female Butch in the history of all animals, but she's a brilliant dog. We sent Alien down to catch and turn the sheep towards the farm, right at that junction, but a terrible mixture of the dog ignoring the calls, his overeager 'eye', plus his lack of experience, led to all the sheep bursting right past him, past the farm, and all the way down the track towards the A66 to Keswick.

We absolutely shit our pants. Annie leapt on a quad bike and screamed down the lane, scooping up Alien as she went. Somehow they managed to get ahead of the sheep, turn them, and send them on their way back up towards the farm. However, as we were now divided as a team, I was left completely alone at the 'junction of calamity' without a dog, but with a 100-metre-wide gap to cover, which, should I have failed, would have led the sheep directly back up onto the miles of open fell we had just managed to bring them all down from.

I shouted, I screamed, I sprinted around and waved my arms like an absolute lunatic, and eventually we did manage to get them heading in the right direction towards the farm. It should've been all over then, but at least half a dozen rogue sheep somehow managed to break free of the group and disappeared completely into a local woodland. Annie, Alien and I still had to go back down the hill and into the woods, before we could finally haul ourselves, red-faced and knackered, back up to the farm and to the rest of the flock.

Derek, naturally, was waiting for us with a broad smile. 'What happened there, then, girls?' he said, chuckling to himself, knowingly.

My journey at Derek's continued, albeit with a big move in the middle. After 30 years of hard graft his tenancy abruptly came to an end on his beloved farm near Keswick. He had been offered another ten years, but he felt he had to surrender the tenancy because he could not get an assurance from the landlord that his daughter Rachel would be considered to continue the tenancy after him. With that, we all moved to a smaller farm near Wigton, a Cumbrian market town wedged between the Lakes and the Solway Coast.

You might be surprised to learn that many farmers do not own the land they work upon. Even if you have a family farm stretching back many generations you are far from secure; especially if the farm is located within one of Britain's National Parks. Stories like Derek's, as sad as they seem, are not even that uncommon, but it still seemed an immense pity that this loyal servant of the land hadn't had his wish to hand over to his daughter respected, after three decades of graft.

Derek taught me the fundamentals of sheep welfare, that I had not grasped in Wales, alongside smaller but equally useful things, like the correct way to tie a sheep's legs together when you need to rescue it on a quad bike, but don't want it to accidentally hoof the accelerator and kill you both. He also took me to my first tup sale – for the big male breeding rams – and he took me to my first sheep sale, for the female ewes. He took me to the National Sheepdog Trials with him, which, as you can imagine, was a bit like walking in behind Caesar as he swept into Rome, and he taught me about what happened when you sent your sheep to slaughter.

One of Derek's biggest gifts to me was just being able to have an association with his great name. From that point forward, having a friend in Derek made getting more work just that little bit easier.

One long summer evening, Rachel and I followed Derek high up onto the very top of Latrigg Fell, a hill that overlooks the town of Keswick and Derwentwater. Derek was preparing to compete in the National Sheepdog Trials that year, and had brought his dogs up for a bit more training, so Rachel and I were able to just sit down on the grass, with a bottle of cider each, and watch the man at work.

The sun was setting over another lake in the District, but I didn't feel like a tourist any more. This National Park, its sheep and its wide-open spaces, were quickly becoming important markers in this new life I was carving out for myself.

The sunset lit up my skin and red hair with the last of its warmth. My body was aching in that deeply satisfying way that only comes with working many hundred hard and happy hours for weeks on end. I could feel my body had started to change too. I had new, functional, muscles that you could never get in any gym. A developing core of deep fitness to hug the deeper well of confidence I was finding within myself, and what I was trying so hard to achieve. I felt my face break into one enormous smile.

Rachel took my picture and I sent it to my dad. 'Welcome to my Office', I typed in the message. The Lake District spread its wings beneath me in every direction and I felt like I really was sitting somewhere close to the top of the world.

A shepherd without a sheepdog is not really a shepherd, though. When my dog Dan departed it was clear at some point that I was still going to need a sheepdog of my own. What was not clear was quite how I might ever find, or afford, that missing piece of the jigsaw.

Chapter Seven
Fraser

'Well, we knew how much you needed and wanted a sheepdog, Hannah. It was your twenty-first birthday coming up, too, so it was a big one. So I rang up Derek.'

It took several years before I thought to ask my mum exactly what had gone on in the build-up to that life-changing afternoon. The initial shock at getting a sheepdog was so great I don't think I'd ever really been able to stop and figure out how on earth it had all happened. Now I'd provoked the memory though, Mum was in full flow.

'Derek told us all about this young dog that you had a really good bond with, and then we had this very townie-to-farmer conversation. Which started with me saying: "So I've bought this giant red bow and I was hoping you wouldn't mind putting it on Fraser, so that when Hannah does open his kennel on her birthday, she will know it's him she is getting as her birthday present?"'

'Oh, Mum!' I burst out laughing. I could just imagine Derek's face, this stoic Scotsman, more used to tying gates and sheep's feet, when faced with the prospect of placing a massive red bow around Fraser's neck.

'Well, Hann,' continued Mum, 'he was pretty horrified. "No, I don't do that to my dogs," was his exact response. I laughed at first because I thought he might be joking. Well, he definitely was not. He was deadly serious!'

I could feel my cheeks burning with embarrassment on behalf of them both. 'So what did you do then, Mum?'

'Well, Hannah, you know me.' She's right: I did know her, and I could see exactly where this was going. 'I posted it to Derek anyway.'

We both started laughing. That was my mum through-and-through.

It was on the actual day of my twenty-first birthday when my parents came to Derek's farm for the first time. Derek and Helen had given me a card and a bottle of something fizzy, and we were all sat around having a cup of tea and enjoying the densest bit of birthday cake in agricultural history, when Derek piped up, 'Hannah, why don't you go show your Mum and Dad how you work the dogs? Give them a taste of what you've learned.' He placed his mug down and waved a hand out towards the yard. 'Tell you what, Fraser's keen for a run out, take them up to his kennel and go get him into the ring.' We pushed back the kitchen chairs and stepped out together.

I was immensely relieved he had picked Fraser. He was by far and away my favourite of all of Derek's brilliant sheepdogs-in-training, and I had loved him from the moment I lay eyes on him. He was a rough, shaggy-coated, tri-coloured collie, with dark eye patches and big ginger-tan cheeks split by a white line that extended down the length of his muzzle, and filled his nose with a beautiful snowy white.

The more we worked and trained together, the more I could just feel we shared this happy energy; an eagerness to get something out of every single bit of the day. There were some of

Derek's dogs that were a pain in the arse to train, but with Fraser the bond was as instant as it was deep. I would've been embarrassed to say it out loud, but I already felt like I had found my soulmate.

The caravan I stayed in was right by his kennel, so I would walk past Fraser every morning. Up he would bounce from behind his door, like a jack-in-the-box, every time he heard my footsteps coming. I would only see his face for a fraction of a second before he dropped back down behind the door, but it was always such a picture of pure unadulterated joy it was impossible for me not to fall for this furry little man. But the thought that this dog, from a sheepdog lineage steeped in Championship achievement and near-perfect performance, could one day be mine . . . well, it was utterly unthinkable. I didn't consider it, even for a single second.

I was definitely feeling tense as we paced across the yard towards his kennel. I had only ever really worked the dogs in front of the people who lived with me down on Derek's farm, and as much faith as I had in Fraser, I really needed it to go well. It was only a few months since the day I had jettisoned the Marine Biology Master's to train in shepherding. I needed to prove to my parents that I hadn't just been wasting my time with this dream. If I'm being entirely honest, the fact that it was Fraser, my best dog, put a little bit of extra pressure on too. A poor performance in the ring would be entirely mine to bear, there was no way I could blame this dog.

I approached his kennel door with Mum and Dad chatting excitedly behind me, and tried to play it cool, but I could feel the nerves twisting like knotted snakes in the pit of my stomach.

I was so singularly focused on what we were going to

execute in the training ring that, by the time I had flung open Fraser's kennel door, I didn't really give the giant red bow round his neck much thought. It must just be Derek being sweet on my birthday, I reasoned. It wasn't until I was actually down on my knees, and giving him a big cuddle, that Mum simply said: 'Hannah, he's yours.'

I looked up at my parents in complete disbelief. 'What?'

'We've got him for you,' said Dad, placing a hand on my shoulder.

I turned to look Fraser in the eye, and burst into heavy, happy tears.

I can remember working him that day, going through all the old routines we had done so many times before, and yet feeling completely differently about Fraser as a dog. I was walking on air. Could he really be mine? The circles went really well. Any sense of pressure had evaporated in that monumental moment. We had an awful long way to go, Fraser and I, but this two-and-a-half-year-old had just completed me.

Fraser was descended from collie royalty. His father was a very famous dog called Mirk, who finished in the Top 10 dogs at both the National and International Sheepdog Trials. It was Mirk Derek had chosen when he was the England captain. Working sheep flowed through Fraser's bloodstream and, as much as I would've felt lucky just to have had any dog, having Fraser meant so much more.

He was young, but he was no beginner. Our training journeys were almost running in parallel. In fact, if anything, Fraser was actually slightly ahead of me. There would be no more delays, and when I did eventually come to leave Derek's, we would both be ready to work.

There were a couple of things that Fraser did need to work on, but one of the massive advantages he had was his exceptional 'eye'. He could walk into a field and would instantly lock in on any sheep. Fraser knew there was nowhere else he would rather be, and our partnership grew in strength rapidly. We were soon able to bring the sheep down from the high fields and into the training ring without any help from Derek. Not an easy task, as the sheep know that they are about to get run around by a bunch of young inexperienced dogs for hours on end. His work in the sheep pens started to improve massively too. Fraser was always a natural at moving sheep in the open fields and around the training ring, but he could be intimidated when we were down in the claustrophobia of a sheep pen, with a hundred angry sheep that all need their doses of minerals and de-worming. The sheep will almost certainly group themselves hard against the back fence; that's when you have to make your dog create space to prise them off and into the pen system. It can be scary for the dog, but it's the only way to isolate individuals to get that job done. He slowly got there, but it was a struggle at first.

The biggest recurring problem we had was that we are both incredibly stubborn and, as Fraser did know a little more than me when working out in open ground, he would often ignore my calls and do whatever he thought was best. What was particularly annoying was that he was actually very often right – his way *was* the most logical and efficient way to move the sheep – but absolutely none of that mattered and was for me to work on myself. What mattered was that Fraser needed to understand that no matter what he thought, even if he was right, if I gave him an instruction, he must follow it to the absolute letter every single time.

To compound the issues we had in those moments, my natural reaction would always be to get more frustrated, angrier and more aggressive with my commands. It quickly transpired that this was the worst possible thing you could do with Fraser. He simply could not take being shouted at. In fact, the crosser I became, the worse he would work, until he would eventually stop working completely.

It was towards the end of my time at Derek's when we were tasked with moving some lambs away from their mums and down into a field of their own. The lambs were old enough to feed themselves now, and their mother ewes needed time to recover before they got ready to raise new lambs all over again. This is always quite a difficult job because, quite obviously, the lambs are going to want to stay near their mothers, but also because the lambs just don't have the experience of being herded on their own yet, so they can panic if you are anything less than 100 per cent efficient. You absolutely do not want to give them any time to think for themselves.

We had managed to separate the lambs off into a field adjacent to the ewes, but what we needed to do next was bring them down the length of the field and off into another field further down the farm. The lambs were all bunched up tightly to the wall that adjoined the field of ewes, bleating to their mums for reassurance, so I needed Fraser to go up to the very top line of that wall, prise them off with a little pressure, and then push them all down the hill, away from me and the ewes, and off into the next field as one.

Well, we started badly, and then it just got worse. No matter what call I gave Fraser he just didn't seem to get that we needed to move them down the field away from us, he just continually and constantly tried to drive them right back towards me and their original position. Pretty soon we were trapped in some Sisyphean nightmare, with the lambs just rotating in one giant circle over and over again, and not getting any closer to actually finishing the job.

The longer it all went on, the poorer Fraser's performance became, and by mid-morning I was absolutely livid with him. The whole job should've been over within ten minutes, but we had somehow only managed to get the flock halfway down the first field. Then the lambs suddenly started to turn and run all the way back to the top of the field again.

'Fraser, COME BY!' I screamed, but he just put his head down and jogged gently to the left, achieving basically nothing, and letting all the lambs head right back past him.

'FRASER!' I was apoplectic by now, shouting, glaring and running up the field towards the lambs that were now powering clean away from us both. That was it. I stomped straight out of the field and went and found Derek. 'I just can't do it. He's not listening to me at all in there!' I raged.

'Are you okay, Hannah?' asked Derek quietly, as we headed back up to the field together.

'Okay? I'm absolutely bloody fuming, Derek. That stubborn bloody dog!' I pointed up to Fraser, who was now just wandering around with his head facing the ground.

Derek pushed open the gate and looked up to see all the lambs still stuck right up by the top wall line. 'Okay, Hann, don't worry. Let's all try again.' He turned. 'Fraser? Come here.'

Oh no. Derek was about to try and use Fraser to do the job. This was highly likely to be extremely embarrassing.

Fraser crept forward to Derek guiltily, looking thoroughly fed up and sorry for himself. 'Now, come by!' he instructed firmly.

Fraser leapt up the hill and just started working. There were momentary wobbles, but Derek's tone never faltered: each command was given with the same sense of clarity and reassurance, and within minutes the lambs were all heading in the correct direction, and were soon down into their new field.

I felt flattened. I had come so far to get to this point, and yet here I was, failing on something that Derek had just made look so basic. 'I know it can be hard, Hannah, but you have to try and be one step ahead of the dog and the sheep at all times. The difference between being just good, and being great, is only your ability to anticipate.'

I still felt like a complete idiot.

Fraser wasn't just being stubborn. He was actually just really upset. Behind his bravado and natural confidence, Fraser hid a deeply sensitive little soul. He take offence easily, and when he does get sad, he just can't work on. Praise him, cuddle him and tell him how amazing he is, and he will walk through fire for you, but shout at him and he will react as if his whole world has been torn apart, and he will just give up altogether.

It probably took six months of hard work for us to reach the understanding that when I raised my voice, it wasn't the end of days, and that actually, in farming, people shout at each other all the time and you shouldn't take it all so devastatingly personally. I actually started taking him to the pub with me too, just so he could get used to noise and shouting, especially by men. Even so, he still has that delicate side to his character, and he can't take it if anyone other than me barks any instructions at him. He'll even run outside if my boyfriend Danny laughs too loudly.

Just as Fraser learned to make some adjustments, I also had to learn not to shout as much and to control my emotions when things did start to go wrong. Derek managing to get Fraser to work those sheep, when I couldn't, might have made me feel pretty small and stupid, but I had to remind myself: I was still learning too. Sometimes we all have to swallow a bit of pride and just get on with it.

There were some things that Fraser was exceptional at. Circling the sheep was one that came naturally, but he also learned to 'catch sheep' at an elite level. 'Catching sheep' is when I'll indicate to Fraser a single sheep that we need to isolate and extract from the herd. This could be for any number of reasons, from it having a sore foot that needs treatment right through to it being on the verge of giving birth, but the beauty of having a dog that can go in and get an individual from a big group is that you don't have to waste any time bringing the entire flock in from the field to grab that one animal yourself.

Even if that sheep manages to give him the slip and run back into the flock, once Fraser has locked on to that sheep he won't ever let it out of his sight until he's finished the job. Catching sheep wasn't something that came to him instinctively; he developed and honed that skill over quite a short space of time at Derek's. His ability to learn quickly, and find ways to adapt, was really important to us because, when we came to leave Derek's, we had no farm of our own to go to. I planned to become a contract shepherd. That meant I was effectively free-lance, and heading to farms when demand dictated they needed skilled operators and extra pairs of hands. That was most likely to be during the seasons of the big sheep gathers from the high fell, and the heavy workload of lambing time. Both had their own distinct sets of demands and skills, all of which we were going to have to master.

We were also going to have to be comfortable working in

fields and farms we might have only set foot and paw on that very day; proving ourselves among dogs and people that would know those places like the backs of their hands. We were going to have to work well together, and fast, most of all because we needed to start building a good reputation, but also because, after a year of working for just food and board, I really needed to start remembering what actual money looked like.

'Well, it's just you and me now, Fraser.'

We walked off Derek's for the last time. Fraser was my team-mate, but he was also my best friend. 'Let's face it,' I said, looking into his deep dark eyes, and laughing, 'you're my only friend here now too.'

My time at Derek's had come to its natural end. After his move to the smaller farm at Wigton, there just wasn't as much work for us to do. He would never have said it, but it was obvious that they really could complete all the jobs without me needing to be there. Predominantly, the work at Derek's was always going to be training sheepdogs anyway, whereas I had always wanted to make my living from sheep farming. I had absolutely loved my time with Derek and his family, and desperately didn't want to outstay my welcome.

From that moment forward Fraser and I were absolutely inseparable. Where I went, he followed. It wasn't like I had some buzzing social life outside of work anyway. Fraser *was* my social life. He was the one I spoke to when no one else was around. He was the one I hung out with in all my free time and, no matter what happened in the fields or farms, good or bad, I always remembered to give him a great big cuddle at the close of the day; usually followed by me either saying 'sorry' or 'you were a bit of a dickhead today'. You could usually tell from his body language which one of those two bedtime sentiments he knew he was going to be receiving. If he'd been naughty, he would come in for that cuddle with his head down and ears pinned right back, but if he was on a high his head would be up, his eyes wild, and his mouth would be wide open, breathing his happy stinky breath all over me.

That first year of working together was a big blur of rough and tumble, high drama and learning on the job, for both of us. A tough lambing season thrust forward into a long summer sheep-gathering with barely a beat in between; but my relationship with Fraser was forged solid in an uncomplicated mix of bloodied grit and determination to succeed. We both wanted to work as much as each other, and that was what always pulled us through eventually.

Our 'graduation' came at the time we had our first big fell gather as a professional partnership, right at the end of our first proper lambing season. We had never gathered on anything that even closely resembled the scale of that job. In Cumbria most of the farms we had worked on had fields extending only about 20 acres, but this was giant, wide-open fell in the very heart of the North Pennines Area of Outstanding Natural Beauty (AONB).

The North Pennines AONB is actually often very unfriendly, very bleak, and always very unforgiving. A mistake here wouldn't result in a shoulder-shrug and slightly red-faced re-retrieve from the far side of a field; the sheep up there could escape into dark and expansive moorlands that might stretch for ten square miles in any given direction. They had everywhere to hide, but Fraser and I had nowhere. If we slipped up on our patch that day then the whole gather would be very badly delayed, or worse: if sheep got left behind, that meant someone's sheep was not getting sheared, someone's sheep was not getting weaned, and someone's sheep was not getting tupped. Bad for welfare, bad for the farm, and very bad for everyone's wages.

There were four farms' worth of sheep totalling well over a thousand ewes on our patch alone that day. If we had failed the chances of us ever being invited back would be absolutely zero; and, trust me when I say this, the farmers round here would never have let us forget it.

To make things so much harder, this particular gather was not like any other in the area. Usually you would start from a high point and work down, sweeping the hillside of sheep as you go. Here, you had to first drive to the top of the fell, then walk all the way to the bottom of the ridgeline and start herding the sheep back up the slopes to the top of the summit ridge. Once there, you had to push the flock down the blind side of the mountain and into sheep pens that were completely hidden from view, up to the point when you were right on top of them. The margin for error was tiny.

We were working within just one vertebrae of a vast mountainous spine that makes up the Pennine backbone of northern England. So, for our gather to work effectively, in this fenceless and wildly exposed landscape, all the farms either side of our four-farm gather also had to gather on exactly the same day too. When it works harmoniously, a chain reaction of gathers sees all the sheep in the whole area pulled off the land as one with, hopefully, only a little bit of 'effing and jeffing' when farms from the neighbouring gathers have to swap a few rogue sheep here and there.

You know when you sit your final exams and you've been

waiting for that day for ages, and you've been imagining it, and you've been trying to prepare yourself for so long, that when it actually does arrive you feel like you've already lived it about a hundred times already? That was what that fell gather felt like when it came around that day. The sense of focus and readiness I felt was so preordained, it was as if the whole thing was happening to someone else entirely. I was watching an actor within my own body, following a strict script to the absolute letter. I had rehearsed it in my mind so many times already, I already knew it wasn't possible for me to make a mistake. It was probably the most 'in the zone' I have ever felt. It was almost transcendental.

I walked Fraser down to the bottom and we started to work back up. I kept him close. Normally I would let him go wide and act a little impulsively, but not today. Today we were both on high alert.

'Anticipate,' I reminded myself of Derek's final lesson, 'the sheep could descend into that ditch, or an over-manoeuvre from simple overconfidence could have them all rushing ahead.' The success of the gather doesn't just depend on one shepherd pulling in their patch without any individual error. We all have to execute in unison, in a long line that must be maintained, and advanced, as one. If a shepherd gets ahead of the shepherd next to them, then they effectively leave the back door wide open in their slipstream. All it then takes is for your neighbour's sheep to run sideways across the fell, slip in behind the patch you've just vacated, and they are pretty much guaranteed to be lost.

'Away,' I commanded Fraser, keeping my tone calm. My sheep could still have scattered in every direction, never mind just sideways, and then I would've had to decide which group to chase. 'Best not let it get to that then, Hannah,' I told myself. 'Anticipate. Anticipate,' I repeated, under my breath.

Fraser was perfect that day, but it wasn't until we summited the ridge that I felt the pressure finally begin to lift, and my body and mind fuse back together. The pens were beneath us, and out to my left and right were all the other shepherds, and all the other sheep, spilling over the long summit ridge and squeezing down into sanctuary together; like grains of sand draining through a giant funnel.

The North Pennines had finally released us from its blackened jaws. Beyond the sheep pens I could see the calm of the Eden Valley and the town of Penrith, beyond that lay the comforting colours and contours of the rolling Lake District and the edges of Ullswater. I could sense it: we were close to home. Passing my 'finals' was now a formality.

It felt truly euphoric walking off the fell with Fraser at the end of that day. The sheep were all successfully divided into the four farms, and I was leading off the ones that corresponded to the farm that marked up their flock like the flag of Japan: white, with a big rounded red spot on the ewe's back.

Fraser and I herded them down a tight trail that split a sea of thick bracken, pressing the sheep into a 100-metre-long white snake that edged its way out towards their farm. I cast my mind back to that pub in Coniston two years ago. Now, I could speak to Fraser in that same shepherd's tongue, I knew how to command a flock of hundreds from behind, and I could have told anyone who would care to listen just why those sheep were on the move that evening. I caught myself for a moment. I could have even moved those sheep through the pub garden myself. I looked ahead at the flock spreading right out before me. I really did know where they were going. Finally, I thought, I knew where I was going too.

Fraser means everything to me. I'm grateful he can't read, because I don't think it would be good for our Leader-Follower dynamic if he ever found out that I actually worship the ground he walks on. I simply can't imagine my life without him and, as much as I know that sad day will eventually come, for now I'm not even thinking about it. Fraser has an awful lot more life to live.

He is more than just a reminder of Nan, though. Yes, he was partly bought with her money, but I believe there is actually a piece of her somewhere carved deep within his soul. Her spirit is found in his kindness. She is there in his softer, more sensitive edges. When he goes, another pathway back to her will go with him, so I remember to celebrate her through the extraordinary amounts of joy he brings me, and all of us, every single day.

Fraser is the only one who has really been with me for every step of my career as a professional farmer. Through all the highs and lows, his presence, his friendship and our partnership has been unfailingly consistent. Even when he's been acting up and we've fallen out, I wouldn't swap him for the world. No matter who comes after Fraser, even if they are better working dogs, he will always be the first. He will forever be my number one.

Chapter Eight
Building the Dream

My family have this funny habit of testing the water on something they are deadly serious about by first making what, to the uninitiated, might seem like a fairly light-hearted comment. For example, you could be sat at the dinner table and someone suddenly pipes up with: 'I wonder what would happen if we took a punt on that crazy new digital photo stuff?' or 'I'd love to try sheep farming, shame I couldn't just completely drop everything, hey?' or 'Imagine if we moved to Cumbria and just bought a farm? I bet that would be fun!'

Be fearful or excited whenever you hear seemingly flippant commentary like this, but be sure something radical is probably about to happen.

Unbeknown to me, Mum and Dad had been thinking about making a significant change for some time. From the moment they had met, they had grafted relentlessly to make their life, and their children's, the very best that it could possibly be. From photography, to digital imagery, to marketing and design, they had built a sound and successful business together from the ground up without loans, debts or handouts. They would be the first to say they had been fortunate, but they also had an absolutely unremitting work ethic and the ability to spot opportunities to diversify and branch out.

Their jobs had seen them move across the corporate and

commercial worlds and into the public health and lifestyle-change sector. It was actually while they were working on campaigns to help people give up smoking and lose weight that they had really started thinking about their own wellbeing in the longer term. All of us children had fled the nest, yet there they were, still working harder than ever. It wasn't like when I was a child, when we'd had nothing, and the fear of not being able to provide probably felt very real; their business was now at a point where they actually had directors working beneath them, and had grown to include contracts and other offices across the United Kingdom. They were still mostly based out of the Wirral, and they could delegate a lot of the day-to-day responsibilities, but having business nationwide meant they were having to travel an awful lot, and their workloads only ever seemed to grow.

Their lives actually changed course on the very same day they'd had such a massive role in changing mine. On my twenty-first birthday at Derek's farm, while watching Fraser and me run through those exercises in the training ring together, Dad had leaned across the rails and asked Derek what he thought were the most common problems when training sheepdogs. 'Auch,' Derek replied, in a thick Scottish burr, 'the problem is always with people, hardly ever the dog.' Dad had laughed. In their years working across so many fields, that had been the main cause of so many issues they had encountered too: people creating problems and barriers for other people who just want to get done what needs to be done.

The collaboration between sheepdog and master repre-sented something of a metaphor in their own minds for a much more straightforward way of getting a message across; a message, they felt, that could be applied in pretty much every aspect of leadership from top to bottom. Soon after, they brought a group of company directors to Derek's, just to learn the sheepdog basics, and found the positive changes in confidence, attitude and outlook of all the participants were quite remarkable.

It didn't take too long for the idea to blossom into something much bigger from that point. All too soon, Mum and Dad started to talk vaguely about 'maybe buying a farmhouse in Cumbria' to start a leadership and behavioural-change course for business professionals and heads of organisations, centred around working with sheepdogs. They weren't about to jack it all in with their marketing and design company, nor was it going to be another lump of work on top of their already massive workloads, but it was a way of starting a new venture, with one eye on eventually winding things down in a part of the country that they already loved.

The whole family were really positive about it all – especially as one of the major motivators was the idea that they could finally imagine a life where they worked a little less – but for me it represented a huge and potentially game-changing step.

I knew Mum and Dad weren't going to be buying some massive commercial farm, but even a small place with a little bit of land presented an enormous opportunity. I was definitely on board to help them establish the leadership programme, and aid all the sheepdog work they had planned, but we all agreed we wanted this new place to be a proper working farm too. Given that only a year before I'd been living in a caravan, without even a functional sheepdog to my name, this change in fortune was scarcely believable in itself; but throw on top the fact I'd grown up in that house at New Ferry, worrying about Mum and Dad spending a fiver on *The Lion King*, and now here we were 18 years later and about to start a farm together, well, the transformation in all our lives really did feel like something out of a Disney film. There was scarcely any time to reflect on any of this, though: the decision was made, and the search for the right place began right away.

I was excited, thrilled and, frankly, a bit terrified but if I had learned anything from this journey so far, it was probably that it was best to commit fully first, and figure out how it could all be done afterwards.

Tucked between the River Eden and the Pennine mountain range, the village of Croglin is little more than a cluster of traditional stone-built farmhouses squeezed around a T-junction. There's a pub, a small post office (which opens just two mornings a week), a beautiful old church, whose site dates right back to the Norman period, and that's about it.

Croglin is probably most well known among outsiders for being the site of what is often described, slightly hopefully, as 'the UK's only genuine account of a vampire attack'. The authenticity of the tale is slightly undermined by the fact that the dates of the supernatural happening seem to fluctuate between two different centuries, and the story focuses on a farmhouse called 'The Croglin Grange' that doesn't appear to have ever existed anywhere in Croglin at all. But, in the spirit of not wanting to let facts get in the way of a good story, here it is.

A young lady named Amelia was lying awake in Croglin Grange one muggy summer's night. She was staring out of her bedroom window, down towards the Croglin church and graveyard, when she saw a humanoid figure sweeping through the darkness towards the farmhouse. As she rushed to lock the door, fierce eyes and a darkened face filled her bedroom window. To her abject horror, the bony fingers of the beast began to scrape at the lead seals from around the perimeter of the locked window frame. Moments later, the pane of glass slipped free, and Amelia, now stupefied with fright, could do absolutely nothing as a repugnant arm swung inside her home, unlocked the window from the inside, and allowed the stuff of vampiric legend to step grimly into her boudoir. Within seconds the vampire's fangs had sunk deep into the pale flesh of her exposed neck, and the 'Legend of Croglin Grange' was born.

Bizarrely, in Augustus's account, after a period of recovery in Switzerland, Amelia actually decided to return to Croglin with her brothers. Naturally, a near-identical scenario soon plays out, but this time her screams alert her brothers, who end up shooting the vampire in the leg as it flees across Croglin's graveyard and disappears into its crypt. At daybreak, the brothers gather a posse, enter the crypt and discover the vampire laying across an open coffin, very much dead from its wounds. However, just to be sure, they threw the vampire on a fire for good measure.

All I'll say is this: I hope it is just a story and not, as some have supposed, a deeply tragic account of how someone who was homeless, mentally ill, or both, was murdered while sleeping rough in the Croglin church grounds some two centuries ago. As I said, 'Croglin Grange' doesn't appear to have ever existed anyway. In fact, there is only one property in the village that truly backs on to that graveyard. It's called Brookside Farm, and it was on the market at a price that seemed too good to be true.

We wanted a farmhouse with a barn, hopefully a shed for livestock, and some land attached. That might sound pretty straightforward in a county where farms often sit shoulder to shoulder in every direction but, in reality, many of those farms are parts of big estates where the farmer is effectively a tenant, not an owner, or they are family-run farms that are passed down exclusively within that farming family for generations. A lot of farms for sale, especially those with any decent slice of land or sizable house, are eye-wateringly expensive second homes for millionaires, or property development projects for the holiday market. Something in our dream was probably going to have to give, whether it was in terms of the size of the land, or the state of the house, or the potential to have your throat removed by a vampire. In the end, it was most definitely the middle of the three.

Croglin's Brookside Farm hadn't been lived in for three decades, and there were parts of it that hadn't been touched in almost 300 years. It had many original features from the early Georgian era: thick stone walls with narrow windows, heavy flagstones on the floor, open fireplaces in every room, beautiful old hardwood beams supporting a slate-lined roof; even a larder with all the meat hooks still hanging overhead.

This place was built to withstand weather and keep people alive when the heavy front door closed shut behind them. Stood in that old kitchen you could have imagined farmers tumbling in from the cold for a few hundred years. The snow and water pouring off thick woollen clothes, frozen fingers stretching out to the fire, a hot broth on the go, bubbling in a thick and soot-blackened ironware pot.

It might have heaved with romance and history, but there was no ignoring that Brookside was in a very bad way indeed. There was a very good reason this place hadn't already been snapped up off the market. The roof was leaking and the damage to the interior wood had left holes in the walls. Whenever it rained the masonry would swell like a sponge, as water soaked into the stone from above and below ground. There was actually one room where rainwater would freely cascade down the wall itself, peeling off the wallpaper and paint as it went, leaving dense, dark patches of damp that were going to take a complete renovation to fix.

My dad was initially absolutely against the idea of buying Brookside, but Mum and I impressed on him just how much potential the place had. 'Potential'. That's a word to watch out for. My parents' friends would come over and say, 'Oh, it's got heaps of potential,' a statement we all now know is actually code for 'You've just bought a complete wreck, what the hell were you idiots even thinking?'

Brookside may have looked like a bargain on paper, but it was going to be the project to end all projects; at the outset, there were probably only a couple of rooms that were barely habitable. I absolutely adored every crumbling, peeling, wind-rattled inch of it, though. The aesthetics of Brookside didn't bother me one bit. Even though it was all a bit tumbledown, it was still ours, and it wasn't as if I wasn't already used to staying in bogging places. Compared to some of the caravans I had slept in over the past few years, this place was The Dorchester Hotel.

I picked out a room with a slightly larger window, slightly less damp, and lime-green walls that had retained most of their paint. I bought a bed, and Mum and Dad fished a mixture of carpet offcuts from a skip for the floor. We were going to have to completely overhaul the place over quite a long time, so there was no point wasting money on any short-term fixes. We picked up a couple of sofas for free and threw some blankets over them to hide the stains. My mum, much to the annoyance of my dad, who had told her not to waste any money on decorating a place we would have to rip apart, took to painting the entire kitchen in a duck-egg blue, just to make it look more homely. Then we got a shower fitted and a little cooker, and I got an old storage heater to plug into a socket, well away from any free-running water.

The bitter cold still found a way to penetrate into my bones, though, especially at night. The thick walls, small windows and all the draughts that entered through the holes meant this house was insulated like a fridge. I swear it was actually colder inside that place than out. It was the sort of house where you would wake up and find frost actually on the *inside* of the windowpane. The people that had lived here before us still lived in Croglin, and would grimly recall the harshest of winters where they all had to take their hot baths while wearing thick woollen jumpers!

I always felt safe and homely there, though. Mum and Dad continued to spend their week working on the Wirral, so I was on my own at Brookside most of the time. I have to admit, though, when it was pitch-black and the dead of the night, and I was having to check on the heavily pregnant ewes during lambing time, the paranormal stories did creep into my consciousness. It wasn't so much the vampire legends that caught up with me, it was the story from the previous incumbents of the time when every single clock in the house had just randomly stopped. It wasn't like they'd had a power cut, either: the old wind-up clocks as well as all the digital and battery-operated ones had all just stopped dead at precisely the same time. It was always enough to put an extra spring in my step and beat in my heart, but I reminded myself that all those haunting stories happened in a time long before us. We have never experienced anything untoward at Brookside; well, not so far, anyway.

What Brookside had in abundance was two priceless things – space and beauty. Outside the house, there was a decent-sized barn – that I would ultimately end up living in for a time while the house was fixed up – and the wreck of a large shed that had been used for sheltering the livestock of the past. That shed was in no way fit for housing any animals, but at least there was something of a footprint of where it once was, so we wouldn't need to reapply for planning permission to get it standing back in place again. Which we would have to do fast, if we were going to realise any farming ambitions within the next year.

We picked up 21 acres of land with the farm. Small in Cumbrian farming terms, where the average size of a farm is over 200, but plenty to keep a small flock of sheep happy. It was upland pasture, not too exposed in terms of height, but I had to be aware that the back wall led to a clean sweep of the North Pennines. A stunning place to farm, a 'pinch yourself' location of real beauty, but I had to be mindful to keep an eye on those mountaintops, as bad weather could soon come tumbling down those slopes, into our fields and onto our sheep at a moment's notice. First, though, I had to get us some sheep.

A farm is nothing without its animals. I knew I wouldn't be spinning us much money out of Brookside any time soon, and that anything we put into the farm or its buildings now was very much an investment for the future, but I still just felt it was important to make a start. Worrying too much would only lead to self-doubt, excuses and long delays. We didn't have a shed, a hurdle, a trough, a hay basket or even a bucket, but we knew we could do it. Build the basics, buy the sheep, and learn on the job.

I actually received Brookside's first couple of sheep as gifts from local farms I'd been working with. Later I would ask to be paid in sheep too; so still not earning any actual money, but building up a flock that would hopefully generate an income further down the line. Ultimately though, ones and twos were not really going to help that much. The sheepdog training aside, the money to be made at Brookside was undoubtedly in the raising of lambs for the meat trade, and if I was going to take that seriously, I knew I had to take my chances in the traditional arena for all farmers looking to expand their live-stock. That, in my area, meant a visit to the famous Penrith sheep auction.

Penrith auction house was a little like Brookside: full of traditional features and not a huge amount of change happen-ing over the last century. Fine, they've got a fairly modern roof and proper lighting in the auction ring, but they only got a card machine last year, and that was only after I had com-plained to the Board. The rituals and customs of the auction are unchanged at Penrith, and when you walk into that auc-tion ring on a big day and see all the old boys sat tight together on wooden benches that have probably moulded to the exact dimensions of their arse-cheeks over the decades, you can easily feel more than a little intimidated by the weight of its history.

I understood roughly how an auction worked from the experiences I'd had when I'd been there to sell and collect other farmers' sheep, but I'd never been there to do business for myself, and I'd certainly never been there completely on my own. I knew I was going to have to keep a firm head on my shoulders if I didn't want to become a walking metaphor for the term 'getting fleeced'.

I was looking for what we call 'ewes with lambs at foot', basically, breeding ewes with that breeding season's lambs as well. The first thing you must do before you even set foot in that auction ring is go and have a look at all the sheep for sale in the main sheds. No easy task. There could be a thousand sheep in there, with dozens of different breeds to choose from. There is no quick way to what you want. You've often got to negotiate your way through a labyrinth of gates, pens and alleyways that make up a vast maze of temporary sheep-holding areas spread right across the shed floor in every direction. Throw on top of that the fact that sheep are being moved around from pen to auction ring constantly, and you soon realise you've also got to be seriously on the ball, and ready to flatten your body against a railing whenever they come sweeping past.

Sheep have the right of way out in that pit, and you don't want to be standing anywhere near them as they come surging through. Sheep startle easily, and many a hapless punter has come unstuck by a leaping animal; worse yet, they are always being led by super-tense farmers, hoping to land top price for their precious flock and plenty ready to issue a stress-relieving and very public dressing-down to any idiots getting in their way.

I tucked into a refuge between two pens and thumbed open the auction catalogue. There lay a blizzard of sheep names, pen numbers, prices and farms. With the blare of the auction Tannoy and the blur of sheep noises and shouting farmers – coupled with the fact that at that point I still secretly thought that all sheep looked more or less the same – well, it may as well have all been written in Ancient Greek. I shut the catalogue firmly and walked out into an open sea of pens and sheep.

*

Sheep farming in Britain is a serious business. We have over 30 million sheep scattered across our shores, accounting for a quarter of Europe's flock, and with around 90 different types of sheep breeds and crossbreeds to choose from we have more varieties here than in any other country on earth. Each breed has its own role to play in providing for the wool and meat markets, as well as their own distinctive look, character and genetic adaptations, tailored to suit the wide array of farming environments locked within our little island.

I wouldn't really know where to begin in describing all of our sheep. Among the rarest and oldest in Britain is the Soay, which originally clung to rough grasses above towering sea-beaten cliffs off Scotland's St Kilda, a small island in the extreme Outer Hebrides. The 5,000-year-old Soay closely resembles the most primitive of sheep, the ancient Mouflon, and, like its island home, it's all rough edges and wild shaggy looks.

One of the craziest-looking sheep is probably the variety known as the Hebridean. Unsurprisingly, given its name, it also originates from the Scottish Islands. We had a small flock once and, I have to be frank, they were probably the most pig-headed sheep I have ever owned. We called them 'The Sons of Satan' as those Hebrideans look like something lifted from a Scandinavian black-metal album cover. They're jet-black, and frequently found with four horns – two thick ones sprouting straight up out of the top of their skulls, and another pairing directly beneath, curling down like brutalist sleeves round the sheep's maniacal face. When one somehow ended up on Brookside's farm roof during a client's sheep-herding exercise, I knew they had to go.

The Herdwick is the iconic sheep of the Lake District. It might have a soft teddy-bear-like face but this is one seriously tough sheep. When held alongside the hardiest high mountain breeds like the Swaledale, Blackface or the Welsh Mountain, the Herdwick almost stands alone in its ability to endure the very worst of Britain's weather. It is great for crossbreeding with lower-lying sheep varieties – when you might want to make that little bit hardier lamb – but they can be very hard work to farm. If they decide they are not up for doing your bidding then, like the Hebridean, they are a serious pain in the arse.

In terms of producing sheep with the thickest meat, it's hard to overlook the dominance of the Texel. Originally from the Netherlands island of Texel, according to the Texel Sheep Society they now account for a quarter of the UK's ram market and 12 per cent of all ewes in Britain. Texels are absolute meat-heads, heavy-set, broad and dense, all characteristics to breed as a stand-alone sheep variety, or to crossbreed into established domestic species for sale to the meat market.

Where I farm in Cumbria it is the North England Mule that dominates. That's a crossbreed of the softer, but very fertile, Bluefaced Leicester ram and the hardier Swaledale, which is also noted for its softer fleeces and powerful maternal instincts. Predominantly the Mules are bred by hill farmers for the farms lower down, but whatever the route, the Mule is ultimately a mix of that mountain toughness, with good fertility, decent wool, a maternal instinct and high-quality meat. It makes them one of the ultimate all-round sheep, and certainly one of the most prolific UK varieties in the north.

Sheep have just as much depth as individual animals as they do as individual breeds. The saying 'to follow like a sheep' has unfairly painted this stereotype of sheep as unintelligent and incapable of acting with any independent thought. This is completely wrong. The reason sheep 'follow' each other is actually because they are extremely gregarious animals that are bonded so tightly to each other socially that the thought of being apart from each other causes them genuine anxiety. It stems from their historic position as a prey species, a throwback to a time when sticking together as a herd offered the only real protection from a grim and grisly death at the claws of a whole host of hungry predators.

Flocks of sheep have leaders; some are escape artists and many are rebels too. Yes, some do 'follow', but just as many are extremely independent and – especially when it comes to facing up to a sheepdog – often very courageous as well. Their co-operation, if you can even call it that, is as much down to you and your relationship with your dog as it is to your understanding of the nature of the sheep you have in front of you. They are absolutely capable of thinking for themselves, and if they decide as a group that they don't want to cross that bridge, enter that pen or place their hooves in that mud, then it is often only your instincts that will save the day.

They have long memories, too. I had a job right down in Hereford once, working as a contractor during lambing time, and I clearly remember this one distinctive-looking sheep that seemed to take a real shine for me. I named her Phillipa. She was a triplet with this enormous head and huge pot-belly, and she came to me every day for a cuddle and a bit of a hand-feed. I reciprocated her friendship warmly and was actually quite sad when it came to saying goodbye. Then, fully a year later, I was invited back to work the same farm. There I was, leaning up against a sheep pen, chatting to the farmer in charge, when I felt a nudge in my side. Incredibly, Phillipa was looking up at me, still bleating for food and my love. The farmer and I were both pretty stunned, but over the years I have seen so many other examples of memory, and a demonstrable emotional intelligence in sheep, that it's hard to deny there is a lot more going on in their mind's eye than we ever like to give them credit for.

Back in the Penrith auction sheds I had settled on a pen of sheep. The guide price didn't look too bad so, pooling together my very limited knowledge, I made a decision. These sheep didn't look too 'woolly'. I wanted them to have a tight thick wool, not something that was shaggy and more likely to get quickly filled with mud and poo. If they weren't too heavily fleeced, it meant their lambs weren't likely to be either, so hopefully, when it later came to selling them for meat, I should get a better price, as the thickness of their actual bodies wouldn't be obscured under a deep mat of hair.

I knew I wanted North England Mules. My reasons for choosing that sheep breed were sound. As I mentioned earlier, the Mule is a brilliant all-rounder that would have every reason to do well at Brookside. Our land was high enough to need something fairly robust, but not so high that I was stuck with just the rangy-looking upland sheep varieties.

Less sound was my fairly unscientific interest in picking the Mules with the best-looking facial markings – the white open-butterfly-like patterning that typically spread like a Rorschach test right across the Mule's blackened face – but with so many sheep to choose from, and little in the way of hard empirical data to fall back on, well, I had to do something to narrow down the vast array of choice in the shed.

With the decision finally made, I picked my way back out of the sheds and headed into the auction ring. I felt like I was a child having their first day at school. There was no friendly face or experienced guide to lean on now. I knew where to get a cup of tea, but here, on the other side of the railings from all the farmers selling, I didn't even know where I was supposed to sit. My hands started to sweat and pretty soon my heart was beating so hard I swear the other punters could see its vibrations pulsing right through my fleece. I was absolutely petrified.

To this day I still struggle to understand exactly what the auctioneer is saying. Although I am getting a little better now, that first time I knew there was a fairly good chance I could either end up paying over the odds, through pure misunderstanding alone, or, worse, not even realise that I was never in the bidding at all.

The rites and lore of the auction room dictate that total discretion is key. Buyers don't want to reveal what they are after, and they certainly don't want to show what they are willing to pay. They'll often feign a complete lack of interest in sheep they are absolutely desperate to have; hundreds of pounds change hands on a quarter-inch twitch of the catalogue in the bidder's hand, or the bidder might be cupping their face in their hands and frowning, and then you'll spot a small extension of their index finger, or just the tiniest of nods. Picking up on those cues is part of the skill of the auctioneer. For the punters wanting to know when to put a sharp elbow into the action, experience learned through the accumulated knowledge of many years visiting the auction ring is vital.

Well, I didn't have years. In fact, given my well-established reputation for wearing my heart on my sleeve, and total lack of a poker face, if anyone I was working for had any intention of buying anything here in the past, then they'd actually ask me to stay well away from the auctioneer, not make any eye contact with him under any circumstances whatsoever and, if I could, sit on my hands too, just for good measure.

As I took up my place to bid on sheep that day I was genuinely worried about exactly what I was supposed to do to make sure my bid was registered at all. It nearly happened to me just last week, bidding on a big breeding male tup. I had to force myself to be so visible in the end that the auctioneer had to break off from his patter completely to say, 'I have got you, Brookside,' before picking it all up again. Mind you, I was glad he did; I had over a thousand pounds resting on that tup's golden balls.

'Ahh-rgh-rgh-rghrhy-rhy-rhym, starting bid at 60, 60, 60, 60. And we have 65. 65. 65. For these beautiful Mules with lambs at foot . . . hmbiderhmbiderhmbider . . . 70, we have 70.'

This was the moment. My Mules were finally in the ring and the bidding was underway. I took a deep breath. Timing is everything. You can't go in too early. A couple of people had started winking and nodding against each other; it felt like it was growing quite fast, but that's the skill of the auctioneer too: making you feel like the pressure is building, willing you to jump in and get sucked into a game where the outcome ultimately means more to you than sticking to your budget.

In spite of the thickness of the auction catalogue, I had somehow managed to roll the whole thing into a tight scroll in all my nervousness. In the end, I wasn't sitting down anywhere either, I was stood up and pressed tight against the railings of the auction ring. My target sheep were within touching distance now, being paraded round the ring by their miserable-looking owner, with the auctioneer presiding over the whole scene, his white coat and arms spread, like Jesus at the Last Supper.

'80, 80, 80, 80, 80. Can I get 85 . . . ?' He broke off and

looked at me directly. I had just caught his eye as I flicked out my catalogue, quite deliberately. He pointed at my chest. 'I have 85, 85, 85, 85, 85. Any advance on 85?'

Shit the bed. I was bloody in. It was about as cool as I think I had ever felt. Looking around the auction ring, fit to burst with pride, I wanted to wave and shout: 'Did you all see that? That's my bid, that is!'

I was about to get those sheep. I could just feel it.

'90 to my right, it's 90. 90. 90. 90.'

And now I was out again. The auctioneer swung back to me: 'Any more? Can I get 95? 95, 95, 95, 95? For these A-class sheep?' *With their beautiful butterfly markings and woolly wool?* I added in my head. 95. Could I stretch to 95? Yes, I could. Nod.

'I have 95. 95. 95. 95. 95. Any more or am I letting these go at 95? Can I get an advance on 95?'

Alright, mate, I'm thinking. *Chill out and drop the hammer.*

'. . . and . . . I'm letting them go for a steal at 95 . . .'

Drop the hammer.

'. . . all done at 95 . . . ?'

Drop the hammer.

'95 . . . It's a steal at 95 . . . going to . . .'

DROP THE FUCKING HAMMER.

Bang. The sweet sound of wooden hammerhead on wooden block.

'. . . Hannah Jackson.' The auctioneer pointed right at me.

Hannah Jackson, I thought to myself: *You have just arrived.*

I had bought 12 Mule 'shearlings', the term for a ewe hovering between its first and second shearing, all with lamb twins. At £95 'per life', it meant I was forking out £285 for each shearling and her twin lambs. A good trade – I was officially a sheep farmer.

That day I led my flock of a dozen ewes onto Brookside was one of the proudest of my life. Against the odds I had got them for a decent price, and the auction experience had absolutely been a positive one. My confidence was sky-high, but it was going to be no good just having lambs and ewes. Even with a further visit to the Penrith auction for a few more ewes, to make any actual commercial gain on the farm I needed all my ewes to breed again, to provide more lambs to sell for meat. For that I needed a tup, a big male breeding ram to impregnate the ewes ahead of the next lambing season.

I had been to tup auctions before with Derek, so I had some sort of an idea of what I was supposed to be looking for in these breeding rams, but I took Mum along for a bit of moral support.

I was after a big beautiful Texel man, with a nice shapely bum, solid back, feet and hips, decent teeth, and the largest, most symmetrical pair of bollocks in all of Cumbria.

Tank the Texel came to Brookside for a half-decent price, but it wasn't until we got him back to the pens that we discovered he was an absolute psychopath. He had this one stubby little horn on the top of his head that he would point and charge at anybody who dared provide him with such inconveniences as food or comfort. It was a wonder he didn't kill me in the pen, but frustratingly, his enthusiasm for violence did not seem to transfer into any energy for sex.

I planned to put Tank to work on the ewes pretty much straight away. I had landed a big job working the entire lambing season for the following year on another farm, so the plan was to impregnate our ewes early in mid-September and get them all 'lambed' (the sheep-farming term for birthing the lambs) by the coming February. That way the main work at Brookside would be done and dusted before the eve of the standard lambing season in early March, and I would be free to work through without any distractions.

To pull this off I was going to attempt a scientific technique called 'sponging' that would bring all my ewes into 'season' (the fertile window of their menstrual cycle) at the same time. It involves inserting a sponge of hormones into the ewes, a little bit like a tampon, for a couple of weeks. Once that's removed, you inject another hormone called PMSG, and then introduce the tup 48 hours later. Everything sounded straightforward on paper, and, up to the point Tank the Texel was introduced, it pretty much was.

The day Tank went to work was one of the hottest mid-September days since records began. There is no margin for error once you've messed with your ewe's menstrual cycle, though; all 40 of my women needed to be 'tupped' (figure it out for yourself) on exactly the same day. I had tried to plan for this potential issue by already borrowing two more tups from a neighbour ahead of time, but as all the big rams wandered around listless and limping in the heat it became pretty clear I had a very serious problem on my hands. None of my boys wanted to perform. Thinking fast, I called in yet more favours and managed to beg another pair of tups from farming friends and, as the day started to cool, the five tups finally set to work.

The clock was ticking and the window was closing fast when, much to my absolute dismay, I realised the randy five were consistently and constantly mounting the same set of ewes over and over again, and flatly ignoring the rest of the flock. I then had to get into the pen with psycho Tank and his mates and physically remove the favoured sheep from their amorous advances. By this point all five tups were panting hard, physically drained and sexually unmotivated. Getting them to do any more work was about as likely as getting Fraser to speak French.

A few months later I had the scanner man, Joe, bring his ultrasound to Brookside. I had hope, but it was a fool's hope. 'Erm,' Joe said, towards the end of the job, with the kind of tone that voices a thousand unspoken words, 'when were they meant to lamb?'

He already knew the plan was February. It was obvious where this was going. 'Erm,' he said again, looking at his ultrasound scan like a fortune-teller peering at a crystal ball for any other outcome, 'I think you might have a quarter of the flock lambing in February, if you're lucky.' Ten ewes. Ten, out of a flock of forty.

To compound matters massively, the vet had previously advised me to leave Tank in with the ewes for another two menstrual cycles; a fairly common practice after you've sponged your flock, just to make sure you've successfully fertilised the whole flock if the tup has missed a couple out. However, as the big day in September had been such a disaster, I had left him in their pens for three. I could never have predicted that only ten ewes would be successfully impregnated that September day, and it had meant that Tank had pretty much gone to work on everyone else at his leisure as their menstrual cycle and fertility windows had settled back down to normal, after all my futile tinkering.

The upshot was, despite my best efforts, I had ewes in

various stages of pregnancy across the whole farm. In the end there were lambs popping out at random intervals all the way through March and into April. I had achieved the actual worst-case scenario, and spent the hardest and most manic lambing season of my entire career running from the contract job on one farm back to Brookside to birth lambs at all hours, all season long.

I swore there and then that I would never make the mistake of playing God with the natural rhythms of a ewe ever again, before thanking God I only had a flock of 40.

There was no time for regrets, though, just lessons, and in spite of everything, I can look back on that first year at Brookside with an enormous amount of pride.

That February I birthed the first lambs on our own farm. One of the warmest Septembers on record gave way to one of the coldest and wettest Februarys and we had only finished rebuilding the wrecked shed a week before the first ewes were due. We hadn't even had time to lay a concrete floor, so Brookside's first sheep were birthed onto straw-covered earth, but they were loved and welcomed as if they were the most treasured sheep in the world. Which to me, and my family, they absolutely were.

My mum was down on the farm the evening we birthed our first triplets. We kneeled together at the rear of the ewe. We already knew she was having triplets thanks to the vet's scans, so we were both prepared. Birthing triplets is never straightforward. The ewe almost always needs some help. It is vital the lambs come into the world one at a time, to avoid damaging their mother, or each other, in the intense squeeze through the birth canal.

I birthed the first lamb successfully, and Mum asked to birth the next two, so I moved away to the head of the ewe and let my mum ease her hand slowly inside. 'Can you feel a head, Mum?' I asked quietly, maintaining the sanctity and calm of the lambing shed. Mum nodded. 'Okay, follow it back till you can feel its shoulder blades, then bring your fingers down its legs till you've got both feet.'

One of the slip-ups it's so easy to make when there is more than one lamb present in the womb is to find the feet of two different lambs and start pulling on them together, making the birth effectively impossible.

'Okay, Hannah, I've got them.' Mum started to pull and out slipped a beautiful little lamb. She cleaned the lamb's mouth and she started bleating right away, but the third and final lamb was much trickier. It had its front legs forward okay, but its head was facing away down its body and towards the back of the ewe's uterus. Not a good position at all. You can't pull a lamb out with its head facing backwards, you'll risk damaging its spinal cord and neck, and it is nearly always incapable of sliding cleanly down the birth canal from that position anyway. Instead you have to try and push that lamb back, deeper into the uterus, and then physically manipulate the head with your hand until it is presenting in the correct forward-facing position.

It can be a really strange sensation when you have to go deep inside a sheep's womb. Lambs have surprisingly long and flexible necks given their size, so as you run your hand in search of the head, it can feel almost as if the neck is going on forever, like a novelty knotted handkerchief up a magician's sleeve.

Space and perspective warp when you are feeling around blindly. I try and close my eyes and focus only on the information being transmitted to my mind from my fingertips, allowing a picture of what is going on inside the ewe to build in my mind without any interference from any other senses.

Mum's whole arm was in past the elbow, first searching for the head, and then trying to turn it; to create that perfect little triangle of two feet and a nose pressed together at the point of delivery. Just as we had seen with Larry the lamb in that field at Coniston.

'I can't do it, Hannah.' Mum's eyes had been closed, but now they were open, staring back at me wildly across the length of the ewe's body. 'You can, Mum,' I encouraged her from the ewe's head. She reclosed her eyes and went in there once more.

As mother-and-daughter bonding moments go, it was definitely an unusual one. In the Western world there are few times where authority truly passes from child to parent. So much of our interactions, especially if there is even a hint of a chance that a life could be on the line, is mediated by authority from outside the family – doctors, vets, surgeons, pilots – yet here we were, together in our shed, with me whispering instructions to my mum that she needed to follow to the letter, and fast.

'I can . . . I can feel its nose on the tip of my finger!' Mum's eyes were screwed tight.

'That's brilliant, Mum, hook your finger round now and gently ease that head forward,' I replied, imagining what my mum could feel, enacting the move she needed to execute with my own fingers. Time was of the essence now; the ewe continued to contract and the heart rate of the third triplet could be falling under the strain of a prolonged birth.

Mum edged her fingers across the face of the in-utero lamb like a rock climber gradually gaining ground on a cliff face. She maintained her grip and successfully flipped the head forward, and, with the logjam suddenly cleared, the lamb slid right out.

Less than a minute later she was stood up with her new sister and brother, suckling from their immensely relieved-looking mother. She wasn't the only one – my mum looked like she had been right through the birth canal too. 'Well done, Mum. I'm really proud of you.'

She looked like she could use a gin or three so we pulled each other up and headed out of the shed. Against all the odds we were really doing it. We had a family farm.

Chapter Nine
Lambing

The lambing season is non-stop chaos. As a contractor I'll often be working on multiple farms from the end of January right through until May. Seven days a week, up to 18 hours a day.

If the ewes are outside, they tend to stick to birthing in daylight hours, but if you're lambing in sheds, with artificial lighting, the ewes will give birth right through night and day. The biggest farm I ever worked on had 3,000 sheep. We had 124 ewes give birth in a single 24-hour period once, but 100 in a day is hardly uncommon on the bigger farms. In a good season I'd expect to get close to witnessing the birth of some 10,000 lambs.

After a big day in the lambing sheds your body aches from ewe-inflicted bruises, and the skin across your knuckles often splits wide open, as the birthing fluids dry hard to your hands. Sleep is a luxury and you eat when you can.

It can be absolute mayhem, like a field hospital in a war zone at times, but this is the chosen life of a contract shepherd, and I wouldn't want it any other way. The adrenaline rush and happy high of seeing life come into the world never leaves you; and, after a quiet Christmas, lambing season is the time contractors can absolutely make hay while the sun shines. Get good at it, find some serious stamina from somewhere, and you can start making some proper money.

With our cash going only one way into Brookside Farm, and after so many months of sheep work experience for just food and board, I wasn't just looking forward to my first full lambing season: I was desperate for it to begin.

Mike and Mary Brough's beautiful farm lay on the northern edge of the Lake District in a pretty village called Hesket Newmarket, a stone's throw from the rushing waters of the River Caldew. It was the first farm I had ever worked on that actually had a tarmac drive, but given I'd swapped the Mini out for a 4x4, my days of channelling my inner Evel Knievel on the morning commute were finally behind me.

The farm itself was really well organised; things made sense and just worked. It was a far cry from what I'd left behind at Brookside. The plan was for me to split my time over the lambing season between Mike and Mary's place, where I actually got to stay inside their farmhouse, in a proper bed, with no crazy peacocks, barking dogs or rising damp, and Brookside, where I was very much continuing to lamb my own flock, look after the sporadic newborns, and help fix up the house whenever I could.

Mike and Mary were warm and very welcoming people. Mary was like a second mum to me, but she also knew how her farm needed to run and would give very clear instructions about what she expected you to get done. Mike was yet another brilliant farmer to be working alongside who, like Derek before him, was a patient teacher, willing to keep my learning journey going, as long as I kept working hard.

They had 1,200 sheep on the farm, as well as a herd of 40 cows; so, a decent-sized farm that was comfortably big enough to require contract workers for the lambing season, especially as both their children had grown up to choose a life outside of agriculture.

The other contractor I was working alongside was the most stereotypical Australian person I had ever met. His name was Geoff, and he was an absolute lunatic. A broad-shouldered, ginger-bearded bloke's bloke, his heavily tattooed forearm bore the 'Stay Ready' slogan of his idol: cage-fighting superstar Conor McGregor.

The first major difference I noticed between myself, the wider British farming landscape, and Geoff's own outlook, was that in his Antipodean culture, sheep farming was about pure numbers. Whether it was the numbers of sheep on the land, or the number of zeros filing into your bank account, there was little room for anything else in Geoff's working world.

Geoff simply could not get his head around how much attention we would give sheep that stood even the slightest chance of not making it. 'Leave that, it's fucking dead, mate,' was his most common refrain whenever we encountered a ewe or lamb in some significant distress. Where he was from, there just wasn't the time. On his farms they would never lamb inside either: 'Good luck doing that with twenty thousand fucking sheep, mate'. Even elementary traditional shepherding was a waste of time, according to Geoff: 'If you farmed as many sheep as us, mate, you wouldn't be wasting your time prancing around in a field with a fucking stick, mate.'

His most regular bone of contention, though, was to question why we weren't farming the largest possible amount of sheep on the biggest available slice of land. His attitude was hardly surprising when you consider some Australian sheep farms are almost the same size as Cumbria in themselves. His home nation is 32 times bigger than the United Kingdom, but with less than half the population; so, bar the actual sheep work, pretty much every aspect of British farming was the polar opposite to what Geoff was used to. Our regulations, our landownership laws, even our sheep breeds and seasons, were completely different; but, in spite of it all, Geoff just got his head down and worked really hard.

Even if I did have to endure the daily 'this is an absolute fucking joke, mate' every time we did anything that wasn't directly income-generating, I don't believe he ever broke a rule of Mike and Mary's. He was just a straightforward 'no bullshit' Australian lad, and most of what he said was very tongue in cheek anyway and, frankly, usually absolutely hilarious.

The more I got to know Geoff, the more I came to discover he actually hid this softer, sweeter side, beneath all his tats and bravado. His father worked on a very remote outback farm, so Geoff was ultimately sent to a boarding school run by nuns for the majority of his youth. It sounds a bit mad, but if you could just ignore all the effing and jeffing, Geoff was unerringly polite, very reasonable and extremely helpful. He would go the extra mile to make sure your day was as easy as it could possibly be, and he always insisted on getting the first round in at the pub; but then he would say something like, 'You know, I once knocked out a camel,' and you'd remember he was always 'loose cannon' first, 'gentleman' second.

During lambing, no one day is the same as the next. No matter how many times I did the job, I never lost that intense feeling of drama and excitement whenever it came to aiding the birth of a lamb at Mike and Mary's place. I always felt happy when a lamb arrived into their sheds well and healthy, and was always very sad if we lost one.

Lambing

Most of the time the births were natural, just like 'Larry the lamb' had been those years before: the perfect little triangle of nose and two front feet presenting at the birth canal, and sliding out under the mother ewe's contractions and pushes. This is by far and away the preferred method of labour and birth, and rarely requires any hands-on assistance from us. Human intervention only comes when it is absolutely necessary. More often than not, the ewe will know exactly what she wants, and where and when she wants to do it. Nonetheless, on a farm with hundreds of pregnant ewes, I would inevitably become accustomed to the full range of challenging birthing positions a lamb could twist itself into.

Some were okay, and just needed a little tweak. For example, if the lamb was presenting backwards, with its head at the rear of the uterus, but still had its hind legs pointing forwards towards the birth canal, you could just grab on to them and allow the ewe to deliver conventionally, just back to front. Sometimes the lamb's knee joints on the front legs could be slightly locked behind the birth canal, in which case you could fairly easily push the lamb back into its mum's uterus a little, unlock its knees and then, again, allow the ewe to deliver as normal.

But there were occasionally harder positions, where intervention needed to be much more direct, more prolonged, and often life-threateningly essential. At Brookside, with the third of the triplets, my mum had experienced a lamb presenting forward, but with its head awkwardly tucked towards the back of the womb; but a much bigger challenge arises if the lamb is presenting in a 'true breech' position. This is when the lamb is turned completely backwards in the womb, and its hind legs are also tucked underneath its body, meaning you have nothing to easily grab and ease forward. Another difficulty comes when twins, triplets and even quadruplets have all their legs intertwined, and you can't easily figure out whose is whose with a cursory rummage. A final, fairly common example in the 'bad birth positions' category is if the ewe is birthing a disproportionately large lamb for her body. All require very sensitive, specific and immediate action, to try to save the lives of both lamb and ewe. Lubrication, manipulation and tender assistance will often bring a positive outcome, or, if there is still time, you can get a vet in to perform an emergency caesarean section. Tragically, there are times when the ewe has already put so much pressure on a young lamb that they simply can't survive. Then the priority becomes removing the dead lamb and saving the life of the ewe. It's a difficult and often traumatic task, which can require a call to the vet too.

The absolute worst moments come when a completely healthy lamb dies from something easily avoidable. For some reason, on very rare occasions, a ewe might not lick their lamb clean post-birth, and a small piece of the birthing sack, left over the lamb's face, causes it to suffocate. Again, another rarity, but it does happen, are the times when, in her haste to cut her lamb free of the umbilical cord, a ewe ends up chewing the cord too close to the lamb's navel and causes a catastrophic hernia. Both cases would only take a micro-second of your attention to swerve completely, and feel heartbreakingly tragic as a result.

As the lambing seasons have racked up over the years, and I have been able to build on my knowledge through the sheer experience of lambing in such huge volumes, I've started to gain a bit of an upper hand in the sheds. One of your greatest tools as an experienced shepherd is your ability to understand whereabouts a ewe is in her labour stage, and whether she needs your

Celebrating my birth with my beloved nan. Just after major surgery she was still the picture of poise and beauty.

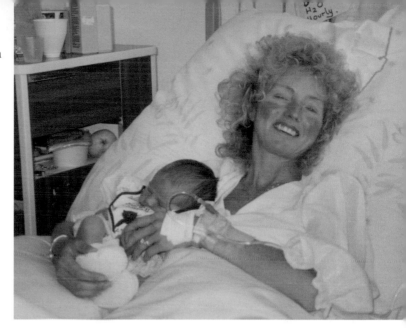

Learning to ride on the Wirral with my amazing dad, Big Stu. Even though my adventures must have terrified my parents at times, they have always been right behind me with a guiding hand.

For as long as I can remember I've wanted to be close to animals. We must have visited every zoo and farm park in the northeast through my urban childhood!

Observing whale behaviour at the Eagle's Claw lookout on a remote Canadian island. For years my only ambition was to study orca for a living.

Just hours after witnessing the moment that changed everything: the birth of "Larry" the lamb, whilst on a family holiday in the Lake District.

The learning curve from townie to professional shepherd was steep, but I knew right away that this was exactly where I was meant to be, and these were the office views I really needed.

My understanding of dogs might have come naturally, but getting them to do exactly what I wanted in the fields was not always easy!

In the early years the fells and lakes of the Lake District became my home from home, as well as my classroom.

Getting Fraser for my 21st birthday was one of the greatest moments of my life. The bond with my right-hand man was instant.

Brookside farm was always going to feel like a "work in progress" at first. But the arrival of the animals soon made it feel like a proper farm to call our own, even surprise ones, like this giant pig!

My mum, Mandy with the first set of triplets we birthed together at Brookside. My Mum has always been a pillar of strength and inspiration throughout my life.

All my animals are special to me, but seeing this girl all grown up really meant something. This is "Hannah", a sheep whose life I saved at birth, strong and healthy down on my friend's farm.

Welcome to my office! The grand sheep gathers on the wide fells of Cumbria became a major source of work as a contract shepherd, but I never lost sight of just how beautiful they were too.

My lowest moment behind the walls on the Estate, where everything was going wrong and a hard lesson needed to be learnt.

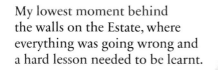

I always try and make time to just sit and watch my sheep, and be with my dogs. Truly, some of the moments I feel most at peace.

CrossFit would become a huge part of my life away from the farm. It brought me new friends, fitness, and, unbeknown to me at the time, in the top right of this photo, even my future husband!

Getting beasted on national television as one of the first female recruits in Channel 4's *SAS: Who Dares Wins* was a major departure from the traditional Cumbrian farming scene, but I discovered new depths of resilience that I've carried with me ever since. Here we are, with the final five recruits in the middle, after an 18-hour long interrogation!

Heading to Australia to work on one of their massive farms taught me a new way to work sheep, but also gave me a deeper love and respect for all our traditions back home.

Danny and I have a habit of getting each other sheep to celebrate big moments in life, but finally being together properly was one of the greatest.

Danny is a full-time farrier by trade, but he instinctively understands sheep farming and sheepdogs. Working Brookside with a truly supportive partner has been a dream come true. This is Drift and Swift our new members of Team Red.

Jim, our rescue horse, has helped heal the wounds of his namesake. The day he laid his head on mine one summer morning, I felt I could see into his kind soul.

The happiest day of my life. Sunset on the Lake District, just after Danny had gone down on one knee and completed me.

Brookside was always going to be a family farm for us all. As a first-generation sheep farmer I don't know where I would be without my family's support. They are everything.

The man, the legend. Fraser at his 10th birthday. The dog that carried me the whole way on my farming journey. He will be a part of my heart forever.

help, purely from the tone and type of voice she is calling out with. Not all ewes are the same – some can really scream at the most routine of births, so you will still have to exercise your judgement – but certain mutters and bleat tones are absolute red flags whenever you hear them, and are critically important early-warning signals. It's a very useful skill to have, especially if you're engaged with helping one ewe while listening to the calls of another; it can really help you prioritise who to work with next. More often than not, though, you'll just be aiding the birth of one ewe and turn to see three more lambs have already popped out behind you; healthily, happily and completely naturally.

It was while at Mike's that I first learned how to pull off one of the hardest lamb rescues. Mike, Geoff and I were all lambing in the sheds on what was already looking like becoming one of the most frantic days of that season, when Mike started shouting for me to come over.

In front of him was an unusually small ewe, with a lamb that was clearly too big for her, wedged hard inside the birth canal. 'Hannah, the lamb in there is absolutely stuck fast. I've tried pushing it back but there's just no way I can move it.' He looked at me seriously. 'I know you've never done this before, but you're the only one in here with hands anywhere near small enough to fit in there.' Mike reached over and picked up a piece of thick cord, 'You are going to need this.'

This is one of the hardest things to do when lambing. Imagine you've stuffed an oversized teddy bear tight inside a long but very narrow drainpipe. You've got to get that bear out, but it's already at the end of your reach, and you can only bring it forward through one end of the pipe. First you have to push it back to find its legs, but you can only use one hand, you can't use your eyes and, before you can even start the manoeuvre, you have to loop a piece of cord around the head of the teddy. Now imagine the drainpipe is moving all the time, the teddy bear is alive, and you've only got a few minutes to complete the entire task. When the method was first described to me by Mike it sounded like absolute madness, but he assured me that if I just took it stage by stage it would definitely work.

I picked up the cord and plaited it through the fingers of my lead hand, just as Mike had demonstrated. I then formed a basic loop: bringing the tag end of the cord down and round with my other hand, and then tying a running knot. Next, I extended my lead hand into the ewe, with the cord wrapped round my fingers. 'She's pretty stressed already, Mike,' I said, just as the ewe called out in the type of wide-eyed, wild shrill that you don't want to be hearing too often.

'Yeah, she's been going for a while now, bless her. But she's never pushing this one out by herself.'

I soon located the lamb's head and started trying to slide my fingertips over its crown, which was stuck tight to the ewe's uterus lining. In hospitals, doctors often work off the principle of 'see one, do one, teach one', but here I was hearing about it, and doing it, for the very first time. The physical pressure in the womb, and mental pressure in the shed, was intense.

'This is actually the tricky bit,' said Mike, once I had somehow managed to create a small space with my little hands. I'm sure he would've laughed at that statement, if it hadn't been quite so serious. 'You are going to need to work that cord out from behind your knuckles now, and slide it behind the ears of the lamb, okay, Hannah?'

I thought to myself, *My hand can barely move in here, Mike, let alone my fingers.* It really was vice tight, but there is still a surprising amount of elasticity to the lining of the uterus; and slowly, so slowly, I started to move that cord forward along the length of my fingers, until it was just resting on the tips of my thumb and little finger.

Gently, I pressed the cord out behind that lamb's ears and eased my hand back, flat across the lamb's head, drawing both sides of the loop down under the lamb's chin as I went. Once I felt my running knot, I successfully worked it into the mouth of the lamb. 'Okay, Mike,' I said, looking up at him, 'I think I've got it.'

The poor ewe was still contracting hard, and trying to push out her lamb the whole time, but it wasn't until I successfully eased that noose tight around the lamb's head that we knew I now had enough of an anchor point to attempt the next move, and get her closer to an actual delivery.

The lamb needed to be pushed further back into the womb to create enough space to pull its legs out from beneath its body, but the use of the cord meant I could absolutely hold that head in the correct place, ensuring I still achieved the perfect birthing triangle once I'd successfully fished the lamb's hooves out from its underbelly.

I gripped the cord tightly with the hand outside the ewe – lives literally depended on it – then I pushed the lamb back hard into the womb with my lead hand. Sliding my hand underneath, I found its legs and pulled them out. The whole thing was a bit like that child's game 'Operation', where you have a plastic man with all his body parts stuffed into little holes on his body, and just a pair of tiny tweezers to retrieve them. Except, failure in Operation meant a buzzer noise and the passing over of the tweezers for your mate to have a go; with this lamb, there could only be me, and if I failed, both lamb and ewe would probably die.

With the legs extracted, I could start to pull the lamb out and put an end to this ewe's ordeal. It needed to come out right away, too. I had kept checking whether the lamb was still moving its jaw, as I'd worked. In such a tight space, the jaw movement was the only vital sign I could look for to indicate that it was even still alive; but that movement had got progressively fainter and, by the end, Mike and I were close to certain the lamb was probably dead.

It was only as I began retrieving my arm from the ewe's womb and birth canal that I realised just how much pain my lead hand and forearm had endured inside. I gritted my teeth, and hoped the lamb would grit its teeth too, but as I withdrew the pain I felt was so explosive I had to press my entire face deep into the rear fleece of the ewe, just to stop myself from screaming out loud. That moment was absolutely do or die; and there was no way on earth that I was going to cry out now and risk startling the mother, right at the very end.

Finally, I felt birthing fluid spill all over my thighs. I pulled my face out of the fleece, and out popped a giant lamb. The ewe was gasping, mouth agape in exhaustion; and there, between her legs, was her child, breathing too.

'Incredible, just incredible.' I looked at Mike. That lamb was hardly the picture of health – it was covered in blood from a huge vaginal tear it had inflicted in the final push – but its mum would recover from that, and the lamb would recover too.

'Hannah, you did an absolutely magnificent job there. Well done. That was one of the hardest I've seen in a very, very long time. You should be very proud of yourself.' Mike and I both took a moment to catch our breath, then parted company to help other not-so-troubled ewes at either ends of the shed.

My small hands have saved many lives since, but that first time with Mike was one of the greatest moments of my career.

It took me a while to come to terms with it, but now, tens of thousands of newborn lambs down the line, I have accepted some sheep and lambs will inevitably die, and that's as much a part of life as it is the job.

At Mike and Mary's, they had a milking machine for lambs without access to their mother's milk. Sometimes it was used by lambs whose mothers had died in birth, but also by lambs whose mums simply couldn't produce enough milk to feed them. This was most often the case when the ewe had birthed twins or triplets, but there were occasions when the ewe simply didn't have a maternal bond with her lamb and would kick it off her teat. As you can imagine, that behaviour was like a red rag to a bull for Geoff, who, without exception, would lobby for the instant summary execution of any such mother; but the milking machine was actually a very efficient, automatic feeder, and a decent enough substitute for the lambs, when it worked out.

It was effectively a mechanical box, with prosthetic teats and a mixer at the back for making the formula milk. It saved me time, as I didn't have to make up bottles (which was, of course, a concept lost on Geoff), and for the first few days I was in my absolute element (it was always going to be me on *that* job). There I would sit, surrounded by these snow-white bouncing lambs, absolutely incredulous that I was now getting paid to do something my nan had paid for me to do throughout my child-hood. However, by the end of the first week, as yet more lambs filed in and my time was getting increasingly stretched, my patience with the process started to grow very thin. 'Come on, little guys, get suckling,' I'd implore, gently trying to tease the fake teat into their tiny mouths, hour after painstaking hour.

By the end of the second week I was getting absolutely fed up with it, and made the mistake of revealing my feelings to Geoff from the other side of the shed. 'See, Hannah!' he shouted over gleefully, elbow deep in a ewe, 'you're already coming round to the 'Stralian way, mate!' After that, I made sure that the job was always filled with a faux merriment. Every lamb matters, Geoff, mate.

When it came to ewes who had lost their lambs, I was introduced to the dark arts of 'adopting'. This is the term given to the process of getting a mother to take on a lamb that is not her own. It wasn't done an awful lot at Mike and Mary's, mostly due to that milking machine, but as I learned the methods it became a skill I found I had a real talent for. An upgrade on just being uniquely useful because I had small hands, at least; and a far better, more natural, solution to the problem of newborn lambs without access to milk.

There are a couple of methods of adoption, but one of the most important things from the outset is that the lamb you try and get the ewe to adopt is of roughly the equivalent age and size to the one she has lost, or to one she already has.

Most ewes have powerful maternal instincts, and some will even show signs of being severely depressed at having lost a child, but they aren't so stupid that they will just take on a new child that is, or was, clearly nothing like their own. But sadly, they aren't clever enough to understand and process that they might not be able to have a child any other way. With ewes, the window for successful adoption is extremely brief, and must involve them not even realising the adopted child wasn't theirs in the first place.

Another reason why size matters is that if the ewe you are getting to adopt a lamb already has a lamb of her own (but you think she has enough milk reserves and maternal capacity to take another), you absolutely do not want to create a competition between the lambs at the teat because, generally, the smaller lamb will end up being out-competed and bullied off by the bigger one. Then you are going to end up having to invest an awful amount of time hand-rearing that lamb right through to the weaning stage anyway. It's also best if the lamb you choose for adoption is hungry, just so they are ready to take to that teat without any delay. Any break in that time-critical bonding moment from birth to breast can alert the ewe to any funny business; and, believe me, 'adopting' is a very funny business indeed.

The two methods we use are loosely called 'wet' or 'dry' adoption. Wet adoption is definitely the less macabre of the two, and certainly the fastest and most efficient. In essence, all you are doing is taking the ewe who has just lost her lamb, or a ewe who is about to give birth to just one lamb (that you think could comfortably rear twins), and covering the lamb you want her to adopt in the ewe's birthing fluids, to make it smell like her own. Everyone has their own unique tricks, but when I do it, I'll mix up a bit of a lamb 'soup' using a big bucket filled with warm water, salt and the lamb I want to be adopted. I'll then birth the ewe's stillborn lamb, or the single live lamb, directly into the bucket, let all the birthing juices mix together, and then present the ewe either with both lambs together as twins, or just with the lamb I want her to adopt. The scent on the adopted lamb is absolutely identical to her own, and the salt means the ewe will keep licking that lamb, bonding naturally, without noticing there was ever a substitution. It works every time.

Sometimes, though, you just don't have that option. You might not have caught the birthing fluid; stillborn lambs are, more often than not, unanticipated; or there might be a delay between the birth of a stillborn and an available lamb for adoption. That's when things get a fair bit more gruesome, as 'dry' adoption involves removing the skin of the ewe's dead lamb and tying it directly to the lamb you want her to adopt.

A lot of farms will cut the fleece off with a knife in a flat rug shape that they then stitch to the adoptive lamb, but when I do it, I try and keep the fleece as intact as possible. I'll even keep the umbilical cord if I can, so that all the vital areas the ewe sniffs, when she's cleaning and bonding with her lamb, remain absolutely consistent, with no gaps around the tail, bum, neck or under the belly.

I cut circles around the deceased lamb's cuffs and collar with a knife, tie its legs to a fence and pull off the skin in one complete piece. It is effectively shaped like a woolly jumper, so when I slip it onto the orphaned lamb there are no openings anywhere, and no way for that new skin to come loose. To the ewe, this new lamb smells exactly the same as the one she's just lost, and again, the bond is usually instantaneous. Later, once I am satisfied that their relationship is firmly established, I'll remove the dead lamb's fleece and dispose of it completely.

It might seem a grisly side of the business, but seeing a ewe bond with a lamb that she is absolutely convinced is her own is a wonderful, joyful moment. You know then that the adopted lamb is going to get brought up the way it should be, not from a machine or from a hand-fed bottle, but from the nourishment of a mother's milk, and all her love. There really is no artificial replacement that can come close to what we already have in nature, but sometimes it might just take a little leg up to get there.

Whenever lambing time comes around, you can be sure someone on my social media will pipe up with a variation of: 'So, why do you spend so much time saving lambs that you are just going to kill anyway?'

There is a very blunt two-word Geoff-esque response to this; or, if you caught him at a good time, he might point out that a lamb's value actually comes at a much later stage of its life, usually some five months in, when they are a very sheepy-looking 40-plus kilos. So, we aren't 'saving lambs' at birth pointlessly because, to survive economically as a farm, you have to get them right through to the equivalent of their adolescence anyway.

To answer like that, though, would be to miss a much bigger and much more important truth. I don't sprint across a barn to save a lamb's life because I'm thinking it will only bring an income to the farm if it makes it to five months, and I don't feed my flock the very best foods and give them the very best life I can afford just because I believe it'll make the tastiest meat in the end. Farming is stressful, it's uncertain, and it's often very poorly paid. The conditions can be exceptionally tough and the hours long. No one is in the sheep-farming industry just for the money, believe me. Most of us love our jobs, and all of us love our animals.

If you are a vegan or a vegetarian, I can accept you have earned a right to criticise the farming and meat industry. I actually respect the choice you have made to not eat meat on ethical grounds and, for you, I'm sure much of the honesty I have already shared in these pages will help solidify your position further. However, what I would ask of you is to fully interrogate the validity of some of the arguments and 'evidence' put forward about farmers from within your community. Especially about how you assume we *all* farm, and what you assume we feel about the animals in our care. Good farmers always have the welfare of their animals front and centre in their minds, every single hour of every single day; and would rather put their businesses, and even their *own* health, on the line than ever compromise the health of their animals. Ultimately though, there is no getting around the fact that we are breeding animals for food; and I understand that the death of any animal for human consumption, no matter how it has lived its life, is upsetting for vegetarians.

Perhaps the greatest compliment I can pay those who choose not to eat meat is that I wish meat-eaters were as conscientious when it comes to what they put on their plate. At one end of the spectrum there is almost a hyper-awareness, where militant vegans amplify instances of poor animal welfare to the highest of platforms. Meanwhile at the other, we have tens of millions of meat consumers who don't even think about their food for a single second.

The vast majority of my country eats meat, and meat-eating worldwide is growing at a rapid rate. Global meat production has quadrupled in the last 50 years, just to keep pace with demand. But so many meat-eaters have little clue where their meat actually comes from, how the animal they are eating was raised, how it was killed and butchered for their plate, and, for quite an alarming number, what that animal even was in the first place.

So there is often an enormous blind spot between those who farm for the meat market and the people who eat the product. I refuse to believe that this is because the consumer just doesn't care. Whenever I have been to schools and spoken publicly about my work, people are always fascinated to learn more, and I've personally found it really empowering when both customer and creator come together in the best interests of basic food and animal welfare standards.

It's hardly surprising that people can be a bit clueless. Everything that can be done has been done to distance the mass market from the meat on their plate. I think farmers could certainly do more to talk about the industry, and just how we do our jobs, but I'm absolutely certain that the way the vast majority of people buy their food today has a huge role to play too.

By the time meat makes it onto a supermarket shelf it is already utterly devoid of all the life and soul of the animal that it came from. All types of animal, and their cuts of meat, are vacuum-sealed behind the same clear plastic film, placed on roughly the same blank tray, stamped with the same tiny lettering that no one ever reads (and, even if they did, only has a very few basic details of the meat's origins anyway) and presented in a featureless straight aisle for the customer to whip through with their trolley.

Meat is basically sold by its sell-by date and its price. The modern shopping experience is about getting the customer in and out as quickly as possible; there's no time for dialogue with a grower, producer or farmer, no time to learn something about the cut of meat, its quality or its origins. Speed and efficiency is the key to making as much money as possible here: get your meat while it's 'three for two', then bugger off down the bog-roll aisle or, better still, buy it with the click of a mouse and never see it lined up on the shelves at all.

The upshot of all this is that the customer is discouraged

from any sort of engagement with, or thought about, the animal itself, and the priority of the supermarket shifts to making sure a bulk product is always available, at the cheapest price it can possibly be. It means there are examples within the industry where farms do not come even close to the same sets of standards that we have at my farm, or the ones I work on. Intensive and industrial farming has become the priority in meeting this 24/7, 365-days-a-year demand; and the thoughtful shopper, and farmer, gets lost in the ether.

Overseas, the rising demands for meat have seen the development of industrial meat-production facilities – I can't even call them farms – that cram their animals together in their thousands, within inadequate pens or fields, and in shocking conditions. Often, they are trapped indoors, unable to see the light of day, stuck eating awful food and stood in their own shit. Many of these animals have to be pumped full of antibiotics, due to the terrible conditions they endure, and injected with hormones to thicken the cuts of meat they'll eventually provide. Even once they've been slaughtered it doesn't stop there: the meat has to be chlorinated or sprayed with ammonia to kill off all the bacteria they've lived with and ingested into their flesh; and later it might even be pumped up with water to make it look artificially plump on the shop's shelves. All that stress, all those chemicals, all that crap, going into their stomachs, then the stomachs of the millions who eat it, and those of their children. It's an absolute disgrace, but the demand and the disconnect from the consumer, and the distributors' focus on low prices and bulk deliveries, absolutely facilitates it.

In the UK we do have basic food standards that must be upheld, especially on our farms, but bizarrely we still import a third of our lamb every year: 100,000 tons of it, in fact, with

Australia and New Zealand accounting for 85 per cent of that. Countries literally on the other side of the planet, at a time when the need to reduce our carbon footprint could hardly be greater. To compound matters, we export a third of our British-reared lamb, too, mostly to France and Germany, because apparently the British consumer prefers a thicker cut than the average British lamb provides. I personally find that hard to believe, and am still yet to hear any customer of mine ever comment on the thickness of a cut. Yet here we are, in a bizarre situation where we are sending meat all around the planet, instead of just enjoying the remarkable home-grown product we have right on our doorstep.

The tragic sum of all this is that many people somehow think very little of throwing meat in the bin. I can't think of any food worse to throw away than the flesh of another animal whose life has been taken so you can eat it. But the trend is growing, especially within the beef industry, with latest estimates showing we dispose of over 30,000 tons of beef every year in the UK alone.

My animals aren't just well looked after. I look after the land that feeds them too. Raising animals for food doesn't always have to add to the environmental burden. Grass is a sensitive product, but care for it with rotational grazing and this natural superfood will feed your sheep unfailingly every single year. Sheep, in moderation, are great for the health of the grass. My fields heave with wild grasses and plants – timothy, clover and chicory – and, probably thanks in part to the monoculture created by the big barley field running along one side of my boundary, they have also become home to wild hares, partridge, buzzard, lapwing, swallows, swifts, field mice, shrews and stoats, scattering and scurrying around all my sheep's hooves. You can't put too many sheep on the grass – it's extensive not intensive farming – but it's worth taking that hit for the health of your sheep. And, my goodness, their meat benefits as a result of all that good stuff.

Even in death I want to give my lambs the best. I make sure I know exactly how and when my animals will be stunned and killed in the abattoir. I have been to the small family-run place that handles the end of one of my lambs' life. I know for a fact that the process is as calming, as clean and as dignified as it possibly can be, and that every bit of that lamb, from its best cuts to its offal, to its bones, will be used for something, somewhere.

But there is absolutely nothing unique about me. I am far from alone in my industry. There are literally hundreds of sheep farmers that produce their lambs in exactly the same way. Quietly, going about their work in all seasons; only for the customer to be later stood in an aisle, reach down and pick out the cut that's been imported from New Zealand and is on a 'buy one get one free' offer. The government could do more to make sure our food is always as green as it can be, and that starts with always supporting the local farmer over any foreign import.

May my industry forgive me for slightly siding with the people who would rather see our business gone completely, but if you eat meat, you *should* bloody well think about where it is from, how it lived and how it died, and then make the *best* choice, ethically and environmentally, that you can possibly afford. Even if it does mean paying a little more, or travelling away from your supermarket to find your local producer, or even eating a little less meat in a week; I absolutely guarantee you, not only will you be doing the right thing for the animals you eat and the environment we all live in, but that piece of meat is going to taste absolutely delicious.

Lambing

For me, as a meat-eater *and* as a sheep farmer, I passionately believe that eating red meat is a central part of a healthy and balanced diet. I find it miraculous that just by helping a sheep be a sheep, in the most stress-free environment I can possibly build, I can produce this amazing, beautiful-tasting end product, of which I am incredibly proud.

So. To answer the question right at the start: I love my job, I love my animals, and I love the meat they produce. I don't see there is any contradiction. I can have all three, and so can you.

Chapter Ten

'You Can't Turn a Scouser into a Shepherd'

For so long I had just been a happy nomad, but now I had Brookside as an actual home I found myself wanting to be there more and more. Even if I did still go away to work on different farms, that place became my constant, my stable, my base; and I poured my dreams for my farming future into its thick stone walls.

Brookside Farm represented a place of coming together for my family. My parents spent all their free time there, and my sisters visited a lot in the early days too. Eventually Holly and her husband James would even buy the house next door.

A year after picking up the keys the place was really starting to come together. My dad, in an ingenious effort to warm the house up a little, had purchased several hundred metres of blue alkathene piping and ran hot water around all the rooms; but as the year gently eased into the warmer months again, the proper building work could finally begin.

I moved into the farm's barn so the roof over the main house could be repaired and a damp course put in. Soon, we were also able to install modern central heating to replace Dad's piping. We lit the old open fires in every room, to draw out all the damp from the stones and bricks, put the farmhouse kitchen back into working order and, with the damp finally gone, we were able to replaster and completely redecorate every room.

Brookside soon started to contribute something towards its debts after that. The first clients of the 'Natural Leaders' leadership and strategy development centre started to file through the barn doors and out into the sheepdog training ring; and, as we rolled into our second lambing season, Brookside's flock of sheep expanded and our first lambs headed for slaughter and sale.

The farm was never just going to be about sheep, though. After my upbringing did you really think I would only ever keep one type of animal? Over the years we have kept ducks, chickens, Highland cows, a beautiful Tamworth pig, and a herd of pygmy goats that turned out to be very accomplished escape artists, constantly either running off into the cemetery or disappearing down the village. They once even managed to board a school bus.

You won't be surprised to learn that my animals had names loaned them either from their life, their looks or my family. The blonde bombshell pygmy goat had to be Margaret, after Nan; but we also had Half-Face the lamb, which had a perfectly black side to its face and was lambed live on Instagram; Pecker the lamb, who lost his tongue when it was pecked out by a magpie; my boyfriend Danny's birthday black sheep Black Betty, with Bam and Pam, her lambs, after the refrain from the legendary rock song 'Black Betty' by the appropriately named 'Ram Jam'; and who could possibly forget Oreo, the black, then white, then black again goat, named for his resemblance to my favourite biscuit. But it was a horse I hadn't named that found a deep, and very soft, spot in my heart.

I could do little about the fact the horse was named Jim. Even if it was a name shared with my nan's partner who had caused so much heartache in our life, the horse had been called

Jim from the day he was born, and there was no way I was going to be changing things around for this dear old soul, who had already lived through so much.

I had never really considered myself a 'horsey' person, but I had always been curious about trying to use a working horse on our farm, just as farmers had traditionally, in a time long before all-terrain vehicles and quad bikes.

Jim was six years old when he was rescued at an auction by my friends Amy and Richard. The second-price bidder was a meat dealer wanting him for mince, but it sounded like he was all skin and bones anyway. His previous owners had not looked after him well at all, and he was in quite a dire state of neglect.

Jim was adopted into their existing herd of horses, but it wasn't until six years later that I brought him to Brookside. I have a flexible arrangement with Amy and Richard, and can technically return Jim at any time, but I can't see that ever happening. They are thrilled I have him here, and the bond I felt with him was instant.

Jim is a beautiful, coal-black, thick-set Dales pony; a native breed of the Yorkshire Dales, historically sent to haul coal in the northern mines and Wales, or work carts. After two months of love, attention and occasional work, I remember us lying down together one long summer's evening and him placing his big black head right on top of mine. It was a highly unusual act of trust for any horse, but something truly special from such a damaged individual.

In that moment, I felt I could see into the depths of Jim's kind soul, and the rush of love I felt was so intense, I thought we could have just melted together into the grass, right there and then.

*

It was during my second year at Brookside that my social media accounts, on Twitter and Facebook initially, and later on Instagram especially, really started to gain some momentum. I had a Twitter profile right when I started out in sheep farming, but back then it was mostly because I wanted to learn more, to see and understand the debates that were going on in farming outside of my area, and to try and build up new contacts. Pretty soon, I realised that if I was going to be noticed at all, then I needed some sort of a name that at least indicated I worked in the industry. It seemed like most of the other farmers were using their farm's name in their handle, but we didn't have Brookside back then, so I started thinking of other options. After I'd gone through all the variations I could think of for the word 'ewe', and had drawn a fat blank, it was actually my mum who came up with 'The Red Shepherdess'. It was perfect. My red hair, the thing people immediately notice, combined with the job I did, and the fact I was a woman in farming.

Some people have since pointed out that the word 'shepherdess' could be seen as disempowering, that if I cared fully about equality I should just be 'The Red Shepherd', but to be brutally honest, in the wider world outside of sheep farming, I'm not entirely convinced the public realise we still actually have shepherds in modern-day Britain, let alone female ones too. For most people the word 'shepherd' only crops up at Christmas time, in the nativity scene alongside Jesus, and then vanishes again for another year. 'The Red Shepherdess': in just three words, it tells you everything you need to know about me, and I believe it was an absolute masterstroke of my mum's.

My biggest asset online has always been the nature of my journey. I was aware that, in a very short space of time, I had gone from being a townie on the Wirral, who had never considered where her food came from, to a shepherd who was absolutely passionate about farming and exactly what it takes to produce meat for the table. It gave me a real sense of empathy and sensitivity for people outside of my industry, and that, I realised, was something quite unusual in my world. As more members of the general public became aware of my story, I became someone who was relatable and approachable, especially when it came to any questions anyone had about sheep farming and agriculture; questions that they might have been embarrassed to ask of an established farmer. I could also answer with straightforward replies that people understood. What I had to say wasn't buried in farmer-speak or confusing acronyms, and I made sure that I was never condescending to anyone. I was once the person asking the basic questions, and I can still remember the day I realised there was a difference between hay and straw.

As well as that, my followers recognised I was still on a great learning journey too. I wasn't someone who had been born into farming and already knew it all. They could follow me and see, almost in real time, my evolution as a farmer: my mistakes and triumphs, the good and the bad moments, the belly laughs and the tears, and, as time passed, the things I was learning and slowly building for myself. Ultimately, I was willing to open myself up and share my life in a way that many other farmers just weren't, and I'm sure it helped endear me to my audience no end. The feedback I got from people was lovely, pretty much right away, so I just kept going; and pretty soon my followers just grew and grew.

At the same time that was all happening, I took on an extra job, pulling pints in the evenings down in our local pub, The Fetherston Arms, known to everyone as 'The Feathers'. As much as the extra money was very welcome, what that job really did was put me in front of all the farmers in my local area. Being the provider of their beer (and very occasionally slipping them a cheeky free pint) probably did more for me, and my career, than any summer spent elbow-deep in their ewes. It gave me time to get to know them and, more importantly, it gave them time to get to know me. The world of Cumbrian sheep farming is very small, and can often feel a little insular too. The only thing worse than no reputation is a bad one, and I wanted to make sure that this relative newcomer to the scene always put her best welly boot forward from the off.

I wanted to make friends too, and pretty quickly found that most of the farmers and farm workers were very welcoming and willing to share advice, once you were embraced into their world as one of them. But, as my experience grew alongside my name, I soon felt myself colliding with a barrier that I knew I was going to have to get around, if I was ever going to maintain any sort of independence or authority as a sheep farmer in my own right. Whether the industry liked to acknowledge it or not, sheep farming is still very much a man's world; and there are things that sometimes happen to farming women that you would never get away with in any other line of work.

My first real taste of this came the same year I started at The Feathers. I had been on this farm for a couple of days, shaping the wool on some sheep to help the client farmer get his prize livestock looking their very best for auction. I'd been doing a good job, making the sheep look presentable and shapely to attract the best possible prices, and even felt confident enough to say to the male farmer that if he needed any more help with anything then he should just let me know.

'Ah, well,' he replied, 'the bigger yows [ewes, in traditional Cumbrian] are coming in next week, so I'll leave that to the lads down the road.' I didn't say anything. What could I say? 'There's more work coming, but the sheep are too big to be handled by a woman like you' was the obvious subtext of his remark. I was absolutely raging. He brought in the tups for a trim of their hooves and walked out.

I immediately made a beeline for the biggest Swaledale beast I could find. *I'll bloody show him*, I thought. I grabbed it by its horns and flipped him onto his arse, right by the barn door, so that when the farmer did come back into the barn, the very first thing he would see was me managing this huge powerful breeding ram's pedicure all on my own.

He walked back in and his jaw hit the floor. 'Fucking hell, lass, well done!' he spluttered. Nothing more needed to be said. I had made my point, but I didn't go back to that farm again.

Another time, I had to go to the vets to pick up an antibiotic order for my sheep, and as the vet handed me the paper bag, he flippantly remarked, 'You picking these up for your dad, then?' I was momentarily knocked back, but I pulled it together enough to point out that: one, my dad isn't a farmer; two, this is my order, look at *my name* on the bag; and three, believe it or not, I am actually the primary farmer at Brookside Farm. Naturally, these extraordinary revelations sent him into a tailspin of: 'Oh, er, sorry . . . I just thought . . . er . . .' until I had backed away from him and his meltdown, and right out of the door.

And how can I forget the time a fence was down on a sheep pen after a fell gather? Fixing that pen (which, it transpired, the lads all knew was broken before we had even started work that day) was a job which would've taken no more than half an hour to fix permanently with a new post, or five minutes to fix with a rope; but no, I was told to stand in the gap while the men went off to divide up all the sheep. That's right: for two hours in the rain, my job was to be the fence.

They might say I was stood closest to the gap, but that's bollocks; and by that point in my career I had more experience with this aspect of sheep farming than most of the boys even up there that day. I was made to do that job because I was seen by them as the weakest link: a woman.

Things haven't changed now I'm established, by the way. Just last week, I went to pick up my lamb cuts that had been prepared post-slaughter. 'Jackson's farm: ten lambs, twenty halves,' I said to the man at the service counter.

'Oh yeah, no problem, where are you from?' he asked. I told him I farmed just over in Croglin. He looked at me quizzically. 'Oh yeah, but *you're* not the farmer though, are you?' he said.

What the actual fuck? I thought. 'Yeah, I am,' I told him straight.

'Really? You're the farmer?' he said breathlessly, as if what I had actually just said was that one of my lamb cuts had burst back to life inside the bag and hopped over his fence out the back. It wasn't like I'd swanned into that place in a flowing ball-gown and full make-up either. I was in rugby shorts, boots and a hoodie. All completely covered in sheep shit.

Just four examples plucked out of hundreds I've had to endure over the years; from very different locations in a farmer's working life, but all typified by their 'everyday sexist' undertones. Don't get me started on the 'It's just a bit of banter, love' brigade. I know the difference between banter in the hills and actual discrimination. There are words we use, and things we will call each other, that you could never get away with in the civilised world. But how is commenting on someone's arse as they walk up a hill ever alright? Or slapping it? Both of which have happened to me in the field before. It wasn't appropriate in offices when it used to happen back in the 1980s, but at least if you did it in an office today you'd be disciplined and almost certainly fired; but no, out in some fields a woman's body is apparently fair game again; and yet, if you dare bring up any of the stuff I've mentioned so far, the anticipated reply is always that it was just either a coincidence, an innocent mistake, or not really sexism at all – 'If you weren't being quite so emotional and sensitive about it' like a 'typical bloody woman', they may as well add.

One in five farmers in this country are female, so you can't just excuse this behaviour by assuming I'm some radical trail-blazer; there are literally thousands of farming women in Britain today, and yet the lack of any sort of ownership from the men who make these comments or my corner of the industry in recognising it is even happening at all is absolutely jaw-dropping.

When the discrimination is absolutely blatant it'll still more than likely be brushed away as a one-off: 'an old-school attitude from a minority of men that are just probably a bit set in their ways', the classic 'you know what them old boys are like, Hannah!' But it isn't just the 'old boys', it is found across all the agricultural arenas I engage in, with men from all ages and backgrounds.

Ask other women in farming, and you would be stunned at how many will say that they don't experience sexism at work (especially if, surprise, surprise, they are being asked the question by a man). I'm sure if you asked those women working in offices, back in the day, where they got their arses slapped as they leant across a photocopier, they would probably downplay it all, too; but that doesn't mean that it is *not* sexist, and it absolutely does not give the industry a pass to say that there isn't really a problem.

Every time I've heard a woman claim that they haven't experienced sexism as a farmer, I then ask them privately if they have ever faced examples similar to mine and they always say 'yes', or that they definitely know other women who have experienced exactly what I am describing. Are they saying they haven't experienced sexism then, because they, like the fellas, believe these things aren't sexist? Or is it actually because they fear being ridiculed, or condescended, or seen as trouble-makers, by the dominant group who are *both* peddling their outmoded views about us *and* acting as judge and jury as to what constitutes sexism or not?

I know being the chief farmer at Brookside, and being effectively self-employed as a contract shepherd, does put me in the privileged position where I can afford to be this direct without it really damaging my career prospects; but I believe, in spite of progress, we are still *at least* a generation behind any other work environment I can think of when it comes to the way women are treated.

The only concession I'll give is that the face-to-face examples aren't the most malicious I've endured in my career. That's reserved for the internet.

As my following online blossomed into the several thousand mark, the confidence of my critics grew too. I don't have a problem taking criticism. There are times when it is fair, and other times when it is just someone presenting an alternative point of view. I might not agree with it, but that stuff is all always welcome; I believe challenging your perspective and being made to think is a really important part of growing and evolving as a person.

However, I started to notice a pattern among one group of people, where it was clear they weren't just offering advice, they were looking to tear me down at every possible opportunity. The first came off the back of a picture of me with a few lambs I had helped birth that day. It read: 'a few pictures of sheep will never turn a Scouser into a shepherd'. It hurt, not just because it was the first direct attack, but because its aim was to pull down so much of what I had achieved so far. It was saying that no matter what you do, or how hard you work, you will always just be some Scouse bird from the city and never taken seriously as a farmer. It might not sound like much, but the unspoken pain of it for me at the time was that it spoke directly to a real fear that I was hiding deep inside. I was still trying to establish myself, and I was still scared people might not take me seriously because of my background.

It didn't seem to matter what I was doing. I posted a picture of me on the way to help a neighbouring farm dig sheep out of a snowdrift five foot deep, and they accused me of being a self-aggrandising chest-beater. The next day we were still there, desperately trying to save the lives of the pregnant ewes, but because I didn't post a picture that time, I was apparently deceiving people. The absolute worst was being accused of building my career by sleeping around, or just generally being a bit of a 'slag'.

It is such an easy accusation for this brand of man to hurl at a woman, particularly as I am sure they know how much that label will really hurt us. 'She's had more pricks than a second-hand pin cushion.' I remember seeing that off the back of so much hate and just bursting straight into tears. I can't actually believe I'm having to write about this but, regardless of anything, if someone chooses to be promiscuous, to have lots of sexual partners, then that is nothing to with anyone – it is *their* private life – but it hurt me personally, because that isn't who I am at all. Having these men publicly say (and Twitter is public) that the only reason I am where I am is through performing sexual favours for other farmers was not only a blatant lie, it absolutely undermined everything I'd built for myself, and just how hard I've had to work, to get myself into this position.

This particular band might've been in an extreme minority, but the biggest issue I had was that they weren't just random anonymous weirdos: they were farmers too, commenting off their farming accounts.

I wrote previously about how I believe farmers should do much more to show the public about farming realities. All of these accounts were doing just that too, but instead of challenging the mistruths we face from outside our industry, they were choosing to spend their energies having a crack at someone who they should've seen as just another ally. I accept some of these men may have private problems, and personal issues, but you can't tell me that some of this isn't also motivated by a deep need to maintain the arcane traditional agricultural status quo that sees only men doing the animal husbandry and physical tasks, and keeps women on the sidelines. Take that snowdrift as just one example. I had male friends going through exactly the same hardships, or also lending helping hands, who were being hailed as 'heroes' by those very same accounts; while I was getting absolutely flame-roasted for posting the same sentiments. The double standards were blatant and, I'm very sad to say, their words cut me far deeper than I wish they ever had.

Our world is a very small one, and it is amazing how malicious rumours can snowball into full-blown, cast-iron facts, despite having absolutely zero merit, credibility or truth – *especially* if that muck is being spread by other colleagues in the industry. What they wrote about me would keep me up at night. I just could not understand how other farmers, who I had never even met, would make such a concerted effort to undermine and bully me from behind their screens. It was the first time I had ever really been bullied in my life and it felt absolutely horrible.

Eventually, their poison had its desired effect. I found myself censoring my posts to avoid their ridicule or criticism and, when that didn't work, I resolved that the only answer was to prove them wrong. For the very first time in my life, I made a decision based on the opinions of other people and not my own, and took a job that would set me up for the most miserable period of my farming career.

Chapter Eleven
Annus Horribilis

I was on my knees trying to fix the steel water trough for what felt like the hundredth time that week, but every time I thought it was fixed, this herd of cows would somehow manage to barge into it and cause it to start leaking all over again; and when I say leak, I mean a gushing fountain of water that would pour out everywhere and empty the cow's entire means of hydration onto the floor in less than a minute. This latest attempt at self-sabotage was a serious one: I needed to find a new stop valve to plug the hole.

It would have been alright if these things were going down at Brookside, or on any of the other farms I'd worked on regularly over the years; but finding a stop valve for that trough, within the maze of the new farm I was working on, was like searching for the metaphorical needle in a haystack. Once I finally did have one, though, I still couldn't get the bloody hole to close properly.

I squeezed the nuts on the valve as hard as I could, but I just couldn't quite get a good enough seal. After about ten minutes of fiddling and straining, it did finally look like the water had stopped dripping from the leaky trough. I got up, turned the water back on and 'bang!' – the water exploded everywhere. All over the floor, all over again.

That little moment marked my breaking point. I held it

together enough to turn the water off at the tap, before I slumped behind a wall and broke down completely.

I have never been someone prone to tears. My ability to put a positive slant on pretty much anything usually carries me through any one of the million adversities you'll inevitably face in farming; but, in that moment, hid behind that wall, I felt about as lost as I'd been since Nan died.

The Lazonby Estate included a large estate house built in the mid-1800s, 1,100 acres of farmland reserved solely for Lazonby business, plus additional land rented out to their own tenant farmers, game-fishing rights to the River Eden, and extra land for seasonal pheasant shooting. It was, in essence, a big, well-established estate, a huge name in my area, and an enormous opportunity for anyone who worked there.

At the same time the online bullying had really started to grind me down, a job came up at the estate, advertising for a farm manager to take on Lazonby's land 'in hand': the 1,100 acres earmarked for their own farming enterprises.

The application process was very competitive; this was, after all, a high-profile farming job with a stable annual salary that would remain consistent through all the feast-and-fallow periods of our agricultural calendar. It was absolutely at the limits of my experience, but what really excited me was their plan to expand their flock of 200 sheep to 800 within the next couple of years. It sounded perfect. Here was a substantial piece of land that was effectively a blank canvas for whoever got the job to pour their own ambition into.

If I was successful with my application, it meant I could really start putting into practice everything I was learning about sheep and sheep farming. As wonderful as Brookside undoubtedly was, the 21 acres we had was always going to be very limiting; and when it came to my contract work, I was forever working on someone else's farm, and only if they were under seasonal pressure. Managing Lazonby's 1,100 acres was the sort of full-time challenge that I felt would really prove my mettle as a farmer. I thought – quite naively – that if I could make a success there, surely it would permanently shut the mouths of anyone that still questioned my ability, my knowledge, or my authority, from that moment forward. I knew already that I was capable of working full-time as a farming contractor, and juggling my responsibilities with the animals at Brookside at the same time, and felt sure I could do the same if I got the Lazonby job too.

In my interview with Serena, who manages her family's estate and would be my boss, I showed exactly how the planned upgrade of their flock could be achieved. I mapped their fields along a strict system of rotational grazing that could maximise the efficiency of their land, and introduced my ideas for a breeding programme that would save them money on buying in sheep, but still allow them to expand the flock to 800 within their two-year time frame. She must have liked what I said, because I soon had a call back to say I had the job. It was late spring in 2017 and, still aged only 24, I was officially the new Lazonby Estate Farm Manager.

For a couple of months everything felt great. I went straight into lambing time and was absolutely in my element. I was working the season I loved, with 200 sheep, and no one telling me what to do. It was May, a late time to lamb by Cumbrian standards, but it was an absolute dream. Beautiful weather had arrived and thick luscious grass grew in abundance for my ewes to graze upon. After that, hay time arrived seamlessly. Again, no real dramas. I had contractors in to manage the job of cutting the grass and baling it up into big hay bales. It wasn't something I'd ever done before, but I was able to lean on the knowledge of the guys who had been here in seasons past, and it all went very smoothly. But that marked the end of the good times I was to have at Lazonby.

Pretty much as soon as hay time was over, I was informed by the estate that their grand plan to expand the flock had been shelved. The vast majority of the 1,100 acres I had been expecting to farm was now to be rented out to other farms for their own grazing. My dreams of expanding their flock, and proving myself, were swept aside in an instant. Things started to unravel pretty quickly from that point forward.

Still exiled behind the wall, I reached inside my fleece and fished out my mobile phone. Tears streamed down both my cheeks, under big snotty sniffs. I was a right mess, and desperately just wanted to hear some kind words. Really, I needed to speak to my mum; but right then she was on holiday with Dad in America, and it was about four o'clock in the morning there.

'Oh, God,' I spluttered, in realisation. 'Look at you now, Hannah. The great big Farm Manager. Bloody crying your eyes out behind a wall and wanting to ring your mum.'

I scrolled down through my contacts until her name and number stared right back at me. My finger hovered over the 'call' symbol.

It wasn't the Lazonby Estate's fault that the moving of that particular goalpost, the ambitions to expand the flock, had been such a crushing disappointment. How were they to know that I was being bullied, and balancing my entire pride and self-worth entirely on making that sheep-flock expansion work?

In my defence, I felt I had made it pretty clear in my interview that my real interest and experience lay with the sheep farming; but I imagine they probably thought I could handle all the other jobs comfortably enough too, and that it wasn't that much of a big deal. Perhaps it wouldn't have been, but things were compounded enormously by a whole series of misfortunes that came in a big run immediately after that revelation. Within a few short weeks I was beginning to feel like the entire job was cursed.

Things began with the flock of sheep we did have suddenly being able to find quite extraordinary ways of escaping the fields where I had left them. It only ever happened when Serena was back on her estate, having been off working at her interior design company in London, or away in America or Ireland, so, it was highly embarrassing and made me look completely incompetent.

It wasn't like they were just turning up in the wrong field, either, they would be out on the main road, wandering around at eight o'clock at night on a Saturday, or in her front garden, munching her flowers, right when she was sitting down to breakfast on a Sunday morning. It was incomprehensible. I had left them behind locked and latched gates. Sheep don't have hands, so what was going on? I had no enemies in my area, but down I would go to retrieve the sheep, and find gates and slide-locks wide open that had definitely been firmly closed before.

Another issue was the state of the grounds. Lazonby Estate felt vast and sprawling anyway; but as I had gone straight into the lambing season in May, I hadn't been able to get a grip of the fairly straightforward task of suppressing the growth of weeds, nettles and brambles with a weedkiller. I was summoned into my boss's car and driven around by her to review the situation. 'What about that one?' she would say, pointing out a clump; 'And that one,' she would repeat, pointing out another. I was burning with embarrassment and humiliation by the end; but it was now high summer, and absolutely the wrong time to be going around hosing down any unsightly nuisance plants.

She hauled in one of the farm managers of yesteryear and, full credit to the old boy, he completely defended me. 'To be fair to Hannah, this is the accumulation of *years* of not being sprayed. It was all well before her time; and if she's full-time lambing in May, how she's supposed to do the weeds and keep the lambs alive?' She was fairly accepting of it after that, but I felt like my card was already indelibly marked.

Next up came the season for barley cutting. I had come to learn by now that Lazonby was actually a mixed farm: hay, barley, cows and – by now – only a few sheep; and that this job really needed someone with a fairly wide breadth of agricultural knowledge and ambition to be an all-rounder. I just didn't have that kind of aspiration, or the knowledge, and there was no one here for me to learn off: I was on my own.

A good analogy for anyone out there that thinks 'farmer' is a catch-all job title for absolutely anyone involved in food production is that taking a sheep farmer and sticking them in a barley field is a bit like asking a mathematics teacher to teach a class on advanced Latin. They might understand the sentiment and structure of the lesson, but they'll be completely lost on the subject itself.

Once I'd blagged my way through the barley harvest, the next job was to plant stubble turnips in the same patch of ground for winter. It's a common way of providing your sheep with a solid fresh vegetable food source in the colder months. It saves you money, as you don't have to buy in much feed, and is straightforward: all you have to do is turn your flock out onto the field for a good munch once the turnips have sprouted. Now, stubble turnips might be a durable vegetable when it comes to braving the elements, but establishing them in a field, it turned out, was not the easiest of tasks for a fledgling grower like me. They are prone to bugs, they can easily be outcompeted and choked back by other plants, or, in my case that season, they can fail for no discernible reason at all.

I'd already sprayed the fields for pests and fertilised the ground before I had sown a single turnip seed, but at the point the muck should've been throwing up some lively green shoots, the field remained brown and lifeless; like it had been hit by a giant mortar shell.

I brought in a field specialist, who shrugged his shoulders and called in an agronomist. Eventually he diagnosed some rare conflict in the soil brought on from planting on top of the former barley crop; but it was well over my head, and certainly over my boss's too.

I felt then like I was stood one side of a giant dam that was holding back a lake filled with shit. At first, just a couple of little leaks sprang up in the dam wall, and it was okay: I had plenty of professional filler. But the leaks just kept coming and soon I was out of filler and desperately stuffing the holes with whatever I could lay my hands on – rags, newspaper, old bits of sponge – until, eventually, out of everything, I was pressed star-shaped against the dam, with every one of my fingers desperately plugging leaks. At that moment the great dam wall splits clean in half and I'm washed away in a massive shit tsunami.

The sheep went from running through locked gates to bursting through broken walls. Then I had a head on collision with another car while in my dad's Ford Ranger, writing off two vehicles in the process. Then I lost the keys to his replacement car in a field. Then my sister rear-ended me in her Peugeot, and completely wrote off her car. Then, Nala, a huntaway sheepdog I owned and absolutely adored, developed tetanus, went blind, lost the use of her legs and had to be put down, all before she was even a year old.

To compound everything massively, I was in a completely unsupportive relationship with another farmer. That year it had become clear that me getting ahead in farming was not on his agenda, and never would be. What I think he wanted was a stereotypical farmer's wife. Someone to cook and clean for him while he worked away on his family's farm. I never felt particularly welcome when we worked together, he certainly didn't seem to value my input at any rate, and the day he effectively told me 'You'll never know as much as me' really should've been the final nail in our coffin. But I stuck it out. I just didn't have the energy for another battle at home, to go with all the ones I was already facing down every single day at work.

Pretty soon, I began to feel really down and withdrawn. I stopped going out, doing things I enjoyed, or seeing any of my friends. I became short-tempered, and the few people I did see told me that I had lost my spark. They were right. My confidence had gone too. The only thing I was sure of was that it wouldn't take much more to push me over the edge.

Still behind the wall. Still hadn't called Mum. But I was talking to Fraser, and not crying as much.

'This place just isn't me at all. I didn't get into farming just to be a maintenance person. Fixing walls, pulling weeds and fixing fecking troughs.' Fraser looked back at me with his big dark eyes. He looked thoroughly miserable too.

I clicked my phone off my mum's contact and thought about opening Twitter. 'Can't do it, Fraser.' I banged the back of my head against the wall in frustration. The hate had been relentless. Getting this job had made absolutely no difference. 'There can't be much to farm managing if you can get the job without any farm management or experience.' That was one of the comments, right from my very first day on the estate. I looked at my dog. 'How could I have been so stupid, to leave behind the contracting jobs I loved, because of a handful of idiots that I'll probably never meet?'

If the leaden weight of my struggles at the Lazonby Estate was sat on one of my shoulders, the trolls were most definitely sitting right on the other, whispering their poison into my ear.

I started to think that maybe they had been right about me all along. That actually, I wasn't ready for this job and probably never would be. Perhaps I wasn't cut out to be a farmer at all? Was I really just this little Scouse townie girl pretending to be a farmer? Was I a fraud?

But I couldn't just quit. Quitting felt like a massive embarrassment for one, but I also knew I wouldn't be able to get another job easily. Quitting meant giving up a stable wage right when we were heading towards winter and the slowest time of year for farm work. You can always top up your wage in a good lambing season, but I felt pretty sure that all the farms I had worked for in the past would have probably replaced me with other people by now too.

I scuffed my boots up in the mud and looked back at my dog. 'Back to just talking to you about my problems again. Haven't done that since we left Derek's, have we, Fraser?'

It really had felt like I was right back where I started, but at least when I had started out, I didn't have a reputation to damage. I was trapped: sticking it out meant being miserable, but quitting might have spelled the end of my whole farming dream. The tears soon began to flow again.

I clicked back onto Mum's number and wrote her a short text message. 'I can't do this any more, Mum. I hate it. I don't even feel like I'm me.' Then I pressed 'send'.

Without doubt, one of the biggest problems I had was my pride. I didn't want anyone to know I was really struggling. I wanted everyone to think I was smashing the farm manager job, and that everything was absolutely fantastic.

Maintaining that façade, that brave face, no matter what happens, is something so many farmers do right across the country, every single day. There is an expectation in this world that you have to be tough and be able to just grit it out as part of the job. Bury your sadness and struggle under a thick coat and thicker skin; and if you can't do that then you're not a real farmer. By which they really mean: you're not a real man.

It is a very masculine and typically masochistic attitude to have, but this job can be tough and unremitting by its nature. There are times when we must grin and bear it. But having that attitude as the default in every single scenario is dangerous and destructive. I'd only been trying to 'hang tough' for just five months, and already felt absolutely exhausted. Other farmers keep up that performance for years on end, until one day they either break, quit, or ask for help; or they end up killing themselves, and everyone is shocked, and says things like: 'I just don't understand. They always seemed and sounded so in control. I had no idea they were struggling.'

The Office for National Statistics shows the risk of suicide among people in the agricultural industry is twice that of the national average. One person dies from suicide every week in farming. Farmers face long hours working alone, often in isolation. They have financial insecurity, where profit can be instantly swept aside by something uncontrollable like the weather, or an unfavourable governmental trade deal, or just some faulty machinery. The line where their work ends and home life begins is often blurred (especially if you live on your farm too), and they often work endless antisocial hours without proper breaks or holidays. You can understand, then, how just these basic conditions of work could all feed a mental health problem, but throw our pride and reluctance to talk on top, and you have a toxic and potentially deadly cocktail; particularly when you consider just how many farmers also have access to lethal tools, machinery and even guns. Farmers are devastatingly effective once they have decided there is no way out.

I have seen people get ridiculed for going on holiday, the suggestion being that they can't be that good at their job, or they can't care enough, if they have time for holidays. It is outrageous and wrong. Whatever we like to believe, farmers don't have a different chemical mix in their heads to everyone else. We need time out and respite too. There are no medals for being the farmer who puts in the most hours and never leaves their farm, just as there is nothing but heartache and sadness for those people who are left behind when the pressures of this job has robbed them of their loved ones.

I was lucky that I had chosen to farm. This job wasn't foisted on me from birth. There was no pressure from a generational farming dynasty that I was expected to maintain, lest I bring shame on the family name. I wanted to be in sheep farming and, until that year on the estate, I had been very fortunate to almost always love my work. I was lucky too that, even in bad times, I was steered forward by a naturally positive outlook, and generally felt enormously grateful that I'd had the opportunity to move into a career that had given me both a sense of pride and a taste of true freedom.

Generally, I lived by the mantra that tomorrow was always a chance to put a bad day right, but on that estate I had fallen into the trap of always waking up miserable and believing that I was about to have a terrible time. It soon becomes a self-fulfilling prophecy, and that bad day became a bad week, became a bad month, became a bad patch, almost became 'my life is falling apart'. Looking back, even though I can't say I had ever descended into a full-blown depression, I was still shocked at just how sad I had become in such a short space of time.

Fortunately, there are some fantastic initiatives in Britain that are aiming to tackle the problem of depression and suicide in farming. Among others, DPJ, established by Emma Picton-Jones in memory of her husband Daniel, who took his life in 2016, diverts its fundraising to training those few people that farmers have to engage with, such as vets or feed representatives, in spotting the signs of someone suffering with a mental health problem. They also run a 24/7 counselling referral service over their dedicated phonelines. Yellow Wellies are another very visible organisation. They deliver training and workshops in both mental and physical health and run the 'Mind Your Head' campaign, specifically targeting depression and anxiety in the industry. They engage directly with a whole swathe of farming organisations, especially at youth level, and even involve the government's Health and Safety Executive to get their messages and training across.

It still feels like we have a very long way to go as an industry, though. The major barrier to progress will always be in finding ways to reduce the stigma around just admitting you have a problem. For me, I've come to realise that a degree of struggle is inevitable on any farm, so admitting you have a problem should never be considered a weakness, it is just part of the job.

I don't believe 30 seconds had gone by before my phone started to ring in my pocket. It was my mum, on her holiday, at four in the morning.

'Are you okay, love?'

Four words I needed to hear, to feel about a million times better. It was a giant transatlantic Mum cuddle.

'I just absolutely hate it here, Mum,' I gushed. 'I've got to leave. I'll do whatever it takes to rebuild myself and my career.' The words started to flow out of me in a rush. The struggles, the worry, the feelings of helplessness. Mum just listened, before telling me not to worry and that we could sort it all out still, that we would write the estate a letter together, that we would make a plan. That the world wasn't ending.

I can't underline enough how much I *hate* making plans, but at that point I was pretty desperate and out of ideas myself anyway. Mum said we only needed to make a short-term plan, just to get out of this dark place satisfactorily and quickly Then we could think about a five-year and a ten-year plan, just to refocus, and stop myself from feeling so impossibly trapped. It was, at the very least, a blueprint to follow towards somewhere else.

'Look at what you've achieved, Hannah. Look at how far you've come from nothing.' I gripped the phone tightly, listening intently, and squeezing the tears back behind my eyelids. She was right. If I could get this far, having never set foot on a farm before, then I could definitely find a way out of this mess.

'When we get off the phone now, take a deep breath and just go home and do something fun. The farm will be okay without you for one afternoon.'

My amazing mum. My nan's daughter. I thought of them both as I left the estate that day. Stoicism and determination in spades, but they both knew when it was time to take a step back and just talk.

I felt a renewed sense of perspective. I would be fine. I'd just lost that connection between myself, my actions and the importance of my own instincts. I will be forever grateful for many of the things my family have done for me, but as the estate disappeared in my rear-view mirror that day, I only wished more people could realise the incredible healing power that lay in just talking to other people about your problems. Rarely are you as alone as you might think.

I knew then that I would be leaving the job, and that it would be the first and last time that a group of people who I had never even met, and certainly didn't care about me, would *ever* have a bearing on what I did with my own life. At least it had only been a year of my life. Generations of men and women have flushed their entire lives down the toilet off the back of trying to please the unpleasables in their lives. The real irony is that now I can actually look back on that period and be really proud of myself for even taking a chance on a job like that. It might not have worked out, but it still took real courage at that point in my career. That was the same courage that I drew on to dust myself down and get going again.

I decided to start right away. That afternoon, I drove all the way to an old friend's house in the Lake District. I had been fortunate to get a job as a contractor on James Rebanks's farm several years previously, and we had since struck up a firm friendship. We shared a real passion for sheep and dis-covered, unlike many farmers, that we both tended to wear our hearts on our sleeves and were capable of being emotion-ally open. We loved a bloody good laugh together too.

I have since grown very close to James, his wonderful wife Helen and his whole family. As a fellow shepherd in the public eye, he has become something of a mentor to me. We both feel the importance of sharing our work on social media, as without the backing of the British public we believe the industry just won't survive.

James is now one of the most famous farmers in Britain. As well as being a first-class Herdwick shepherd, he is a best-selling author, and has cultivated a large and dedicated following online. Of all the people I knew, he was probably the next best placed (after Mum) to lend me some advice.

That day, he gave me the almightiest bear hug the second I arrived at his farm. I told him about the job, the trolls and my deep concerns for my future in farming. His advice was simple: 'Fuck 'em all. I'll have you back to work here any time.'

Well, that's a start, I thought to myself. *Should probably give the boyfriend the boot too.*

The Red Shepherdess was on the comeback.

Chapter Twelve
This Shepherd's Life

It was summer in Cumbria again and I was back in a happy place. It was shearing season. The time when all the sheep must have their thick fleeces removed so they can be comfortable in the fields through the warmer months. The sale of their wool used to bring bonus profits for farms, but the price of wool has slumped to such a low in recent years that, once you've paid your shearing team, you're lucky if you break even. Still, it must be done for the welfare of your flock, so there I was, on Mains Farm, working an all-new contract.

The Lazonby Estate accepted my resignation without any qualms, and, in the end, I hadn't had to grovel for any of my old contracts back either. There was, naturally, a lot of piss-taking to endure, along the lines of 'Oh, having to grace us common folk with your presence again, your Royal Highness!', which I gratefully took on the chin, under the obvious exterior shield of 'Yeah, and I'll bet you've all fucking missed me, haven't you!'

My failure at Lazonby was never going to be an awkward 'elephant in the room' in my little corner of Cumbria, and I was really glad that it wasn't. By a wonderful stroke of good fortune, James Broom, my sister Holly's husband, had taken on many of my contracts while I had been at Lazonby. After I left, he successfully applied for my old job as farm manager, which allowed me to slip right back into my previous line of work, as if I hadn't even left in the first place. James is still at the Lazonby Estate today and, I have to say, he *is* the right person for that job, and absolutely loves it. I'll even help him out down there from time to time.

Mains Farm was a beautiful place to be. Our shearing patch overlooked the pure running waters of the River Eden, breaking into clear views of Lazonby village, with the Lazonby Estate hidden just around the bend, a little further upstream. After only a few months away, I couldn't have felt further removed from that place. I bore absolutely no ill will towards the estate or Serena. It just hadn't been the right job for me, and I wasn't the right employee for her.

The farm buildings and walls of Mains were made of a coppery-looking sandstone. Its spacious barns were entered through dramatic high archways, tall enough for a double-decker bus to pass clean through, and the whole place was well organised, well run and spotlessly clean. Well, spotless for a farm, anyway.

There was a campsite attached, so we could always hear the distant sounds of children chasing each other around the fields, music being played, and the heady scent of barbecued meat and beer drifting through the air and directly into our nostrils while we worked. Whenever we needed to move sheep from the fields to the farm's yards, the campers would all rush to line up and take photos of us. You had to watch they didn't distract the dogs or get in the way, but I felt a strange sense of déjà vu whenever my eyes met with those tourists. I was that person once, down in the pub at Coniston, in what already felt like a different lifetime.

We were set up for shearing outside, with a large shearing

trailer lifted off the ground, ready for the shearer to take up their position. It was a three-person job: one to guide the sheep up the trailer's ramp and into a 'race' (a narrow corridor), where the shearer would grab hold of the sheep, do their job with an electric clipper, before the third person (me on this occasion) gathered the freshly shorn fleece and folded it up into the giant British Wool bags, ready for sending on for pricing. When the whole enterprise is running fluidly, with a good shearer up on that trailer, you can comfortably get through 45 sheep in an hour.

I was back working as part of a team. No more isolated days with just Fraser and my own thoughts for company, and I was already feeling a whole lot better about everything. Negativity feeds off negativity, cannibalising itself until it has grown into an all-consuming beast in your head. As soon as I was back bouncing off other people, I felt my sense of perspective return, and all those bad thoughts from the previous year drifted away into the fresh air.

It was seriously hot and sweaty work, though. It was the day after the first major storm of the summer, so the air was thick with humidity. Perspiration streamed down my arms as I scooped up the wool, but I certainly could not complain. At least I was in shorts. The shearer was dressed in a heavy uniform that included a thick shearing vest and even thicker, double-denim shearing trousers. It offered him some protection if his clipper slipped in his hand, but it meant we all had to be mindful of his body temperature, and made sure he fired fluids into himself in the rare moments he wasn't engaged with shearing one whole sheep every 80 seconds.

I soon fell into the sort of hypnotic rhythm that you can only achieve with any task that involves lots of repetition. Gently, I allowed my mind the space to reflect on all that had happened over the past two years.

I felt I knew myself better than ever now. Getting right back to where I had been before, so quickly, had proven I had every right to believe I really did belong in this industry. If all those farms hadn't wanted to have worked with me as a contractor again, they could scarcely have had a better opportunity to blow me out. It was, after all, me who had upped sticks and left them for the estate.

My focus on succeeding in that job, to silence those critics, had been way off the mark. Change wasn't going to come over-night, and I now knew the best thing I could do was to focus on doing the best job I could in the corner of the industry I absolutely loved: sheep farming and contract shepherding.

There are many industries that embolden themselves with the idea that only 'special' people can take part. The truth is, though, there are actually very few jobs that you can't find a way into, if you really want to. Farming can feel like an old boys' club at times and, when you take the world of traditional sheep farming in isolation, it feels even *more* impenetrable, even *more* traditional and tribal, than almost any other agricultural discipline I can think of. Particularly if you're an outsider with-out a farming background, and *especially* if you're a woman too. But I had eventually been embraced here through my own hard work and dedication. Now, I just needed to keep going.

I think one of the biggest problems is that certain men can feel threatened when women move in to what they feel are their roles. What I want to say to those men is that I am not here to change things and undermine your traditions. I don't want to tear down your systems and replace you, nor do I want any special recognition or treatment. I'm not looking for a fight either. All women like me really want is inclusion and a level playing field. Straightforward equal opportunities to do what we want to do, and live our lives the way we want to live them. We are not striding into your world to crush your manhood or step on your toes. I just want to be able to stand in a field next to you, and be considered as just another shepherd.

The irony is that in much of the wider world beyond our shores, agriculture is dominated by women. Farming is by far and away the biggest income generator for females in the developing world, and yet women own less than 20 per cent of the land they work, they earn substantially less than men (despite working more hours worldwide) and they have considerably less access to the funding, the technology and the market places where their product and graft gets turned into cash.

In the United Kingdom, I mentioned before how just one farmer in five is female; but that number is growing steadily, and there are now twice as many women as men studying agricultural subjects in higher education. Prospects for women in this job are undoubtedly improving – we even have the first female president, the brilliant Minette Batters, at the head of the National Farmers Union – but there is still this prevailing myth that it is really the men who are doing the actual hard work, and that women are present in a sort of 'back seat' role, making the soup and ensuring the men's wellies are warm in the morning. It completely ignores the huge number of women that are out there, like me, running every aspect of their farms and farming lives.

Old habits die hard. The reputation, tradition and inheritance line of farming is still steeped in handing down from fathers to sons in a patriarchy extended and defended over hundreds of years. I know excellent and experienced female farmers who have been made to stand aside for their far less competent younger brothers, and I also know of tenant farms that have been passed to entirely new families because the next in line to acquire the right to the land was a woman.

In Scotland, the government have set up a 'Women in Agriculture' task force specifically to give women the same level of access to opportunities as men. In Wales, the Welsh government have a 'Women in Agriculture' forum, specifically to allow women to inform the government on agricultural policy; but in England, we still have virtually nothing at a governmental policy level specifically to support farming women as a group; and it is nothing short of a disgrace.

Whenever I turn up at our local farmers' meetings, I'll often be the only woman in a room of ten men, regardless of how many female farm workers we have in my area. Now, I'll have the craic with anyone, but there are women who will feel intimidated in those meetings. More to the point, though, when I question where the female farmers are, the answer that usually comes back is a variation of: 'At home looking after the kids, someone has to.' Yes, someone does, but why don't *you* bloody do it once in a while so your partner in life and work can come here and get her opinions on all of our futures laid out in this room? It's *never* the woman and *always* the man. It is why they do have women-only groups and forums in Scotland and Wales, and why we must have them in England. It isn't discrimination against men, it's about making absolutely sure there is always a place for women, literally and metaphorically, at all the decision-making tables.

I told my boyfriend it was over. That it just wasn't working. That we were moving in two completely directions. That we had very different opinions of the roles of men and women in farming. I couldn't be 'The Farmer's Wife' for him, and he couldn't cope with the idea of me being in the fields and farmyards full-time.

The farmer's wife. There is absolutely nothing wrong with being one. I just resent all the stereotypes that title immediately presents. It is the humble woman alone indoors, while the heroic man does all the animal work. She is the cook, the cleaner, the solo mother; without aspirations or apparent ability to be a 'proper' farmer herself, she is forever dependent on him as the 'breadwinner'. These falsehoods around the 'farmer's wife' are so pervasive it is *still* the dominant image of what most people think of when they picture a woman married to a farmer, or even women in farming generally.

When women are considered in the field, it is usually radical solo figureheads like Hannah Hauxwell that immediately spring to mind: the legendary spinster and lone Dales farmer who lived without electricity and running water in an often quite-brutal state of near perpetual hermitage right through to the 1980s. Hannah became nationally famous after a documentary of her extraordinary life, *Too Long a Winter*, was aired, and is unquestionably a farming legend and icon. But I guarantee you now that if she had ever married, it would be her husband taking all the credit for *her* radical farming ways. More to the point, in spite of all her undoubted achievements and deserved recognition, Hannah Hauxwell will forever be an outlier who bucked a trend, without starting one. Just look at any children's book with a farmer in it today. It's not going to be a woman you see, it's going to be some flannel-shirted, flat-cap-wearing man. The best we can apparently achieve is to be married to a male farmer, instead of running the entire show ourselves.

'Behind every good farmer is his wife,' is the other tired, cliché phrase you might hear as the truly pathetic attempt to give a 'pat on the head' to all those females who are married in agriculture. I wouldn't mind if the farmer's wife raised the more realistic image of a woman who is in a paid business partnership with her husband: the person who is absolutely managing the farm too, splitting the animal husbandry, doing the paperwork, generating income and, more often than not, looking after the children and doing all the domestic chores on top as well, but it isn't, is it? 'Behind'? What do you actually mean by 'behind?' The farmer's wife is absolutely front and centre in helping to make a farm function, and that's the best we can do for them? To place them 'behind' the man?

Some may still say there isn't a sexist undertone in all this, and to that, all I'll say is, well, I'm still yet to meet a 'Farmer's Husband'.

I chucked another fleece into the bag and headed back to the shearing station. The shearer was just finishing another smooth job on a very compliant-looking ewe. I've always thought elite-level shearers look like they are ballroom dancing with the sheep they shear. Mine was wearing a pair of traditional shearing moccasins on his feet too. They resemble leather taco shells with laces, but are surprisingly comfortable, and allow the wearer to slide their feet around the floor of the shearing platform without slipping. The shoes are deliberately thin, letting the shearer's feet act like an extra pair of hands, feeling the movements of the sheep and letting them glide quietly under the sheep's body to hook into them in certain ways – underneath the back leg, or under its rear – without ever making the sheep feel like it is being manhandled or forced.

The smoother your dance, the more relaxed your dance partner. As soon as a shearer tenses up, or tries to manipulate a sheep with muscle, the whole routine breaks down and it quickly becomes a fight. Our man today was absolutely on point. His gently persuasive dance moves meant all his work was progressing as a collaboration, and not a battle of wills. We were making really good time.

I moved faster than my shearer, and stole a moment just to watch him work. It was the same moves every time. He started on one of the sheep's hind legs, slid the electronic blades under its wool by the ankle, and worked from that point up towards its rump. He then delivered a few blows up one of its sides and headed up onto the sheep's neckline. At that point he split the wool, worked backwards from the neck, down the other side of the sheep, before eventually finishing at the ankle of the remaining back leg. The wool came off the sheep's body as one complete fleece, and fell in a pile by the shearer's feet, ready for me to collect it up again.

There are many romantic ways to describe the correct way to wrap a fleece, but, in truth, you wrap them in exactly the same way that your fish and chips are wrapped at the local chip shop. There might be some variation on the theme depending on the sheep breeds (some fleeces are thicker and denser than others; others you fold inside out) but that day it was the Mules, so commercial and conventional.

I pulled the whole fleece off the trailer and spread it out with the thickest, most luxuriant side of the sheepskin facing down. It's a rough rectangle, so I began by pulling the sides in and rolling from just one end. I repeated the process, pulling the sides in and rolling it until the whole thing was tightly bound in one thick tubular shape. By beginning with the best bit of wool facing down, you should find that, when you've finished, that piece is always presenting on the outside of the roll. In that way you are advertising the fleece at its best; meaning that when it comes to the point the wool is graded, you should get the nod for the best possible price.

In contrast to lambing time, where your hands are dry and the skin fractures, the lanolin that secretes off the skin of a sheep coats your fingers and palms as you roll, and gives you the softest, most beautiful-feeling hands imaginable. A good day's rolling is better for your hands than any soak in the most expensive hand creams you could imagine. I guarantee you now that every farmer, male or female, notices it and – whether they dare admit it or not – absolutely loves it.

With the fleece successfully rolled, I'd toss it into the British Wool bag, a carrier so big I'd actually pack down the fleeces by climbing into it and jumping up and down until they were all compressed tightly at the bottom. I can fit at least 60 Mule skins in a single sack before I close it all up, totalling over 100 kilos of beautiful sheepskin.

Shearing is timeless. Take away the electric clipper, replace it with the old hand tools, and you are looking at a scene that has taken place for as long as these lands have held domestic sheep. The folding and throwing of the fleece, the heat and the banter, would be as recognisable today as they would have been several centuries before. It is an iconic and ancient rite of the shepherd's year. But the shepherd's life has changed. Modernisation and globalisation have brought new markets, fresh comforts and seriously efficient technology; as well as new stressors and a level of scrutiny that would've been utterly unheard of in times past.

When I was born, less than half a per cent of the world's

population were engaging with the internet, and the idea of 'social media' was more of a philosophical debate than an actual tangible platform for communication between people. Fast-forward to the time of writing and you'll find nearly 5 billion people active online, representing almost 70 per cent of the planet. Computer technology and the internet has been the fastest-growing sector of engineering in the history of humanity, and companies like Facebook, Google, Instagram and Twitter all have the audience and influence to change the world at the click of a mouse.

It is far from all good. The almost uncontrollable speed of growth has lent an opportunity to people who want to spread extremism, fake news or just to bully others, with complete liberty and very few consequences. I still think there is a lot more good than bad on social media – I wouldn't use it to talk about my work if I didn't – but there is a festering boil out there that needs lancing, and those companies shouldn't just be leaving it up to their own users to police what is right, and what is wrong.

When it came to all the trolls online, I only wish I could have spoken to myself a few years ago and said that it doesn't matter what someone says or thinks about you on the internet. It isn't real life, and it's a game you can't win. It didn't make a difference whether I had the farm management job or not. Before I had the job, I was being slagged off. I took the job, I got slagged off. I left, I got slagged off.

The trolls are not going to go away, but the trick is seeing it for what it is: a small, probably bitter and jealous, possibly unwell, minority of people that choose to hate, against an overwhelmingly positive majority that enjoy what I post. I'm not going to come offline and give them what they want, and they are never going to stop this Scouser from being a shepherd.

I believe it is more important than ever to share the stories of our farming life far and wide, and today, as a woman in farming, I know I have to find that strength to keep proving that we can do it, against whatever odds and barriers they put in our way. Ironically, a lot of the strength I get to keep going, in forging my own path forward, does actually come from my followers. I am so lucky to get so much love and encouragement from that other, far bigger, group of people I may never yet meet; and I feel immensely grateful to every single person who has taken the time to give my work a 'like', leave a nice message, or ask me a question about farming.

One of the most useful things about having a large, long-term following today is that I have built up genuine loyalty over all the years I've been posting. I recognise I'm now in the privileged position to talk about some of the knottier issues and harsher realities of farming (many of which are detailed in these pages) as well as dispel a fair few myths. If I didn't have that following, and weighed in online with a video of how to 'dry adopt' from a dead lamb, from a standing start, I'm pretty sure most people would be appalled; but my followers, who have been with me long enough to buy into my morals and values as a farmer, trust me to show them farming reality, warts and all. I hope by being as transparent as I can online, I am helping bring my industry closer to a mainstream audience and, in that way, closing the knowledge gap between an often very defensive sheep-farming world and the wider public who consume our products. I want my profile to help build allies in our food production and consumption, to celebrate the highs and the lows, and, I hope, arm my followers with the reasons why we should all value what British farmers do every single day.

No matter what anyone says, you can't 'just ignore' online

hate. As hard as I try not to, I will still get upset by it from time to time, but not nearly as much as I did. I've learned now that they aren't suddenly going to stop because one day I've 'proven myself' to them. That day will never ever come. For these people their social media accounts exist as sport, where they gain cruel entertainment and a twisted pleasure from attacking other people.

The people that run those big platforms, the bosses of Facebook, Twitter and Instagram, all know they could and should do more to combat bullying, but much more should be done within my own world too. I'm established, and have enough friends within the industry, that people will defend me when they see me under attack online, but it would make a huge difference if the bigger farming accounts and organisations weighed in to shame these idiots into putting down their keyboards too. I'm not the only woman in farming who receives abuse, and I know that they all see it, so let's join forces and try and combat it together.

We reached the end of a long day at Mains Farm and the whole shearing team sat down on top of the big British Wool bags, the physical expression of all our hard work. It was already late evening, and stomachs were rumbling, when the farmer strode out of the farmhouse with a huge earthenware pot filled with lamb curry and a bucket of ice-cold beers. Gleefully and gratefully, we scooped the curry into bowls, and placed all the beer bottles into a steel cattle trough to keep them cold.

The food and drink didn't touch the sides that day, we just inhaled it into our stomachs and laid our weary bodies back on the bags, allowing the evening to close in all around us.

Women aren't just these weakling humans that are always

going to be a burden when it comes to physical tasks in the field. Remember, after all, it was women who took to farming during both World Wars, keeping food on the country's table while the men were away on the frontlines. I do accept the fact that women generally aren't as physically strong as men, but that does not mean we are always incapable when it comes to any task that involves muscle. Also, it is actually possible to find new techniques and alternative ways to get a job done, without just employing pure brawn. In fact, there are occasions when a more thoughtful, less physical, approach actually works better anyway.

That evening, you could have just asked the dancing shearer who was lying down next to me, but I can't forget the lessons of Derek Scrimgeour in teaching me the value of quiet poise and presence when it comes to the sheep gather. That going in all guns blazing can actually be counterproductive to the task, putting off the dog and heightening the anxiety of the sheep.

On the gather, I've come to realise that sheep want time to look around and process the scenario they are faced with. If they are put under immense pressure from the off, they tend to go into a chaotic and panicked state. They stop thinking straight, start jumping over walls and fences, and will scatter easily. Pretty soon, the task of getting them off a fell has become a lot harder, simply because it has all been turned into a battle of wills, muscles or egos. A better way of gathering is to try and place yourself in the mind's-eye of the sheep from the offset; consider them and what they might be thinking; and then guide, not force, them towards what you need them to do.

Probably the greatest place for a change in mindset is at lambing time. This is a deeply instinctual and natural moment between the ewe and its lamb, when the ability to take a step back, empathise and analyse, is hyper-critical. So often, though, I've seen blokes just go barging into this quite sacred scenario, and physically impose themselves on the situation; grabbing ewes forcibly to milk the lambs, crashing around and pushing machinery, gates and livestock about, in a misguided attempt to get results as quickly as possible.

It almost always causes intense apprehension among the ewes in labour, which, if you observe any mammal, or ask any woman who has gone through childbirth, is the absolute worst thing you can do. Instead, I'll try and spend as much time watching as I possibly can, and if I do need to intervene, I will hope to have at least worked out the most discreet method possible.

I'll enter the pen and stoop down, bringing myself immediately to the eye-level of the ewe, making myself as non-threatening as possible, and I'll even talk to them, quietly and softly, about what I'm doing and why. I know they can't understand the words, but it keeps all parties calm, and soothes the mum no end. In all my experience, watching all kinds of people at lambing time, this is the way I've learned to get the best out of the sheep, and I believe it is the most time-efficient approach too.

Just as the idea that women are physically less strong than men is a generalisation and not a rule, neither is it always true that farming men are never capable of being considerate, nurturing or, dare I say, more feminine in their approach. There is a huge amount of room for nuance, a fact especially true when it comes to discussing sexism in the industry.

I know for a fact that there are some very open-minded male farmers – Derek Scrimgeour and Mike Brough are two mentioned in this book, for example – and that not all male farmers hold prejudices against women in farming; but it would be completely disingenuous for me to pretend it just doesn't exist, or that it isn't a problem that I've personally encountered on many occasions in the past.

I am never going to change the opinions of the hard-line misogynist. They are radicalised in their opinions against women, and have already made their judgements on what or who they think I am; but I do believe they are in an extreme minority. The people I want to reach are the women thinking of having a go at farming, but are nervous about what the industry might think, and those farmers who might never think they were sexist, but actually hold prejudicial opinions that they don't even realise are there. Those people who, when I reveal that 'No, I actually am *the* farmer,' stutter backwards in disbelief. Those men are fully capable of change, and it is a change that, when it does happen, is only going to make British farming stronger, and feel more inclusive, for everyone.

There are many fantastic female farmers out there that are finally getting airtime, column inches and growing large audiences on social media platforms. Public opinion is changing, so surely the time has now come for us to look at ourselves as an industry and ask, honestly, what sort of image do we want to project? I want the image of a British farmer to simply be that of a person who is proudly employed in feeding the nation. I don't think that is too much to ask.

Chapter Thirteen
Fraser's Peak

The relationship between shepherd and sheepdog would ordinarily be a bit like the one between parent and child, leading ultimately to the point that you, as a parent of sorts, let go of the reins and allow your four-legged protégé to enter the world, and start acting independently of you. However, as much as they are free from your direct instruction in that moment, you hope your teachings, messages and values will still be guiding all their subconscious actions towards the behaviours that remain the correct ones for working out wide on the fell.

With Fraser, it was always going to be a little different. Fraser was my first dog, acquired at the very start of my journey in farming. We have learned and developed together over the years, with neither one of us quite having complete mastery over the other in terms of our knowledge and experience. I was not his parent, just as he was not my child; however, the day would still come when Fraser would need to break away. He was a talented dog, and I could feel him pushing the boundaries every time we headed off to a big gather.

If he were the yo-yo and I was the string, I knew at some point I would have to think about letting that control slide off the end of my fingertip and just hope, as the yo-yo rolled away from me on the floor, that my string would wrap and weave its way intractably inside his body. I wouldn't know if he was truly ready until it was all actually happening out on the fell, by which point it would already be too late to pull back if we had got the timing all wrong.

It is the ultimate leap of faith, and some dogs, and some masters, can never let their string go. There are sheepdogs and shepherds who are forever interned on the easier stretches of fell; where the sheep are always in sight, the dogs are within a safe calling distance, and the margin for error is wide. That's okay, though. Some dogs just can't be trusted to go it alone. Maybe they don't have the instinctual, emotional or intellectual where-withal to act completely independently, and maybe they never will. They might freeze in the moment or, worse, go rogue, start chasing or being aggressive towards the sheep, if left completely to their own devices. Sometimes, it's the confidence or experience of the shepherd that holds the partnership back though. Maybe the human doesn't feel quite ready to take the leap, where bad weather conditions, unsteady ground or vast expansive fell conspire to steal your self-confidence.

The thing is, on a big gather you always need dogs and people doing the less technical work on the easier ground anyway. Not all of the terrain will be complicated to move sheep across, but they must all be gathered; and, as long as you maintain the correct distance and hold your patch of the line, then you can't go too far wrong.

Out at the wing tips of a gather, though – the extreme edges, where the cord of the net gets drawn in along an invisible line – well, that's where you need the best dogs, and the shepherds with the most skills and experience. Even then, the moment may never arise where that shepherd needs to send a dog out beyond the reach of their calls, but you need to know that the dog up there has it within them to do that job, if necessary, because out wide, they are the backstop, the final barrier between the sheep staying safely ensconced within the agricultural world, or disappearing off into a deadly wilderness.

I had been gathering with Fraser for a few years, pulling sheep off hard fells as a contractor right across Cumbria. We had executed several technically challenging jobs and, on fells where I had the confidence of a little prior experience, we had begun to really push the envelope: pulling sheep from distances that sat on the very edge of my calling distance. When Fraser was a quarter of a mile away, my words would sound to him like a murmur, whispers on the wind, but they were enough to keep him honest, and focused on the job. But I couldn't help but wonder . . . what would he do if my calls weren't there at all?

There was no real way of telling if we were ready for the next step until the moment arose, but there was equally no knowing when or if that test would ever actually come. Our test did come though; it was on the Great Dodd, an iconic horseshoe-shaped ridge that summits at its centre at 857 metres. It is the highest patch of fell to the north of Sticks Pass, tucked away in the north-eastern quarter of the Lake District, and, in terms of the places we gathered, it was up there with the toughest of the lot.

I had gathered the Great Dodd before. The horseshoe ridgeline sits high in the sky above you, as if this U-shaped shoe is being held aloft on the fingertips of an elderly giant. Those wizened, fractured fingers run away from the horseshoe as becks and gullies, hanging down along the slopes, before they conjoin to meet the giant's wrist, forearm, and then the elbow, which presents itself as the top of a lower secondary summit, before a further descent brings you to a great valley floor. The landscape and sense of exposure up there is massive. When you are stood at the bottom of Great Dodd, in this enormous expansive amphitheatre, you realise there is absolutely no way of calling across to your fellow shepherds from one side to another; if, in good weather, you are blessed enough to even be able to see an escaping sheep on another's patch, you have to resort to mobile phones to find a fix. Fundamentally, though, you are on your own until that whole area has been swept, and you won't have close human contact again until all those sheep have travelled down the fingers, forearm, and onto the giant's elbow; the meeting point for everyone to begin the drive away from that place, for yet another season.

Shoddy was going to be up there, of course. The King of the Fell, the professional fell gatherer, the man who has gathered more sheep on fells like this than the entire population of Cumbria has had hot dinners. He was going to take the summit of the ridgeline, pushing sheep down the slopes and hemming them onto the inside of the horseshoe, but the technical positions directly beneath him were vacant. He was the first to say it should be Fraser and me down there.

'But, Shoddy,' I stammered, slightly taken aback, 'I don't even know where I'm going.'

'You'll fucking work it out, lass,' he snapped back, before immediately heading off.

Cheers, Shoddy. That's the end of the procedural instruction then, I thought.

There were plenty of other shepherds and dogs on the gather (there had to be whenever we worked Great Dodd) but we were missing a couple of the most experienced team members, and that gap had to be filled with the next best replacement. I just hadn't quite expected that to be us.

The Great Dodd feels naturally intimidating, but we had learned the ropes on its beginners' slopes and I knew we could take more responsibility. The extreme right-hand side, though? That was an area on the outer edge of the horseshoe, the exact opposite side to where we normally worked. Out there, there was nothing to stop stray sheep that might ignore our persuasions from disappearing off down the backside of the giant's hand and into the unfettered wild.

You may think that gathering somewhere new is no big deal, especially for a contractor like me who is constantly turning up on new farms, fields and fells, but the Great Dodd presented so many unique problems that you never really knew which way the sheep would run under pressure. As Derek would repeat, anticipation is the key to a successful gather, but this was like anticipating the exact run of a rolling marble across a paving slab that's just been smashed with a sledgehammer. The mixed landscape up there, the bracken, the gorges, the hanging mini-valleys, waterfalls, water traps, and added to all that the grey boulders that were identical in colour, size and shape to the Swaledale sheep we were moving, all added up to a labyrinthine level of obstacles and potential fuck-ups, stretched across miles and miles of mountainside that were only even visible at all *if* the weather held.

As we moved into position that day, the one doubt that still gnawed at the back of my head with Fraser was that stubborn and hard-headed side to his character. He'd had a keen taste for freedom from the first day we met – he was never going to be shy of leaving my calls behind if the moment came – but it had taken an absolute age to train him to listen to *my* instructions in those moments when he felt his way was best. I feared that he might slip right back into those bad habits when my calls weren't there to suppress his own decision-making, particularly that excruciating habit he had of driving the sheep back towards me, when he should be moving them all away in the direction of whatever holding pen was the destination of the day.

In the end, the decision to let him go came and went in the time it took for a single ewe and lamb to lift up their heads.

Everything had been going fine. Yes, the terrain was undoubtedly more complex, but we could cope. We started to move from our end of the horseshoe's ridge, sweeping up the sheep, squeezing them, and moving them on, feeling out the ground and getting a sense for how the field of play might influence our movements as we went, but as soon as my eyes met with that mother and ewe, I knew they were going to break free of the careful mould we were making.

A Swaledale ewe and lamb, with black faces, white eye-patches and white noses, like inverse panda bears; their heads had bolted up immediately from the grass they were eating. They had considered us for a micro-second, and then sprinted off in completely the wrong direction.

'Ahhhh. Hellfire.' I looked up to the top of the horseshoe ridge; Shoddy was up there, but those sheep were angling well behind him. *He won't miss this,* I thought. Shoddy has an eagle's instinct for movements on the fell, especially irregular ones. *I'll be in even bigger trouble for not trying to bring those sheep back than I will be if I just leave them and concentrate on bringing the rest of this flock along successfully.*

That ewe and lamb weren't for slowing down either, they were making a bolt for freedom on the far-right flank, the wide-open spaces of what remained of the Lake District. We were the back of the backstop. There was no one else even close to them, and they were making a very good fist of leaving us for dust. This was it then. Hold my position here with the rest of the flock out in front, and send Fraser to catch those rebels, all on his own.

I pushed in front of Fraser to start the move. Normally he would be no further back than the line of my heel, but I wanted to let him know that what we were about to attempt was the greatest flanking outrun of his career.

'Awaaaaaaaaaaaaaaaaaaay,' I elongated the 'a' sound as far as my breath would carry it, showing verbally as well as physically that he needed to go extremely wide. 'Go out! Go out! Go out!' I called after him as he ran, fixing in his mind that this was the time for a straight sprint, there was no point cutting in: there were no other sheep to find; the only thing that mattered was that pair, and we were gambling the whole house on him making it.

Those initial calls were still so important. Fraser couldn't see the rogue sheep from his diminutive height, only I could, so it was an act of extreme trust on his part that he kept up that pursuit across the highly uneven terrain, through at least two waterfalls and several gullies, based only on the eyesight of his master. As the sheep continued to flee, I became aware that the moment would soon come where my calls would cease to be of any help to Fraser. My eyes would no longer be a part of his body, and my voice would vacate from his head completely. Soon, it would be all up to him.

The moment I knew he was acting independently came when he was merely a tiny black dot on the landscape, closing in on the pair of white spots on the ridgeline, like an ant-sized arcade Pac-Man, trapped within the hardest play setting imaginable. He was well out of calling range and the sheep had made it above another waterfall, almost to the top of the ridge, almost out of sight, almost gone for good. That was the point he could've easily turned back. He could see the sheep now, but he was as isolated as he had ever been in his working life. The pressure in his head to give up, turn back and come down without them would've overwhelmed him at any earlier stage of his career, but not that day. That was Fraser's peak. It was his time.

In the final moments before the sheep could have escaped, Fraser caught them. He turned them from the outstretched finger ends of the Great Dodd giant, reversed them back down the slope from the summit ridge, and started hauling them down towards the flock, towards the giant's elbow, and not just back towards me. It was the moment that all of our training, all of our working, the hours of calls and whistles, the repeated gathers, collective successes – and learnings from our collective failures – came together in one singular beautiful expression of our relationship. It was as close to perfection as we have ever achieved.

I contained my excitement until all the shepherds, dogs and sheep had made it to the giant's elbow. Don't get me wrong: I wanted to shout, 'Look at my dog!' at the top of my lungs when he was up there, godlike in the heavens above, but it took a lot more work and serious concentration to progress the gather, and the flock, across this hanging landscape where sheep and dogs were constantly dropping out of view. When we did make it, though, I embraced Fraser with the warmest cuddle I could muster.

'I cannot believe you just did that, Fraser!' I rubbed his weary muscles. You could see in his eyes that he was very happy, but he was absolutely knackered too. His tongue lolled from his mouth and his shoulders heaved in heavy breaths, forcing oxygen into his body to recover. He was a supreme athlete before today, but he had just proven he was a first-class sheep-dog too. 'I cannot believe you did that, Fraser! Good boy! Bloody good boy!' I was repeating myself, as I saw Shoddy approach from the edge of my eye-line.

As much as Shoddy would amplify our errors, both big and small, with a vintage ear-quaking blast of 'Yer fucking useless dog!', he was also capable of recognising good performances too.

'Fucking hell, lass!' he enthused, brimming with excitement at what he had witnessed, 'your dog did bloody well there, hey!'

In a land where compliments are hard-won (in fact, they are almost seen as a weakness), where a gentle extension of a single finger from the handlebars of a quad bike, in simple acknow-ledgement of your presence, is the equivalent of a gushing embrace in the civilian world, praise for you is to be very grate-fully received; well, inwardly at least.

I would have been deliriously happy with or without a com-pliment coming our way from Shoddy, but to get one from him, from a man who knows so much, whose opinion on fell gather-ing I hold in such high esteem, it really did mean the world to me. 'Your dog did alright there,' was the nicest thing Shoddy had ever said to me, or anyone else that I've been around for that matter.

'You feeling alright, Shoddy? Actually saying something nice to me for once, are you?' I retorted, which translates in normal-speak as: 'Thank you, Shoddy, that's wonderful of you to have noticed. We really appreciate it,' which he countered with:

'Well, it won't fucking last, will it!' accompanied by a cheeky wink.

*

What Fraser did that day was make a statement. He laid down the marker for where we were as a partnership, and for where I was as a shepherd. Even if I went out on the next day and made an absolute pig's ear of a gather, we had just proven ourselves capable at the highest of levels. It gave me confidence, but it also meant that I didn't feel that need to battle for belonging any more. From that day forward, Fraser and I were always among the first to be asked onto the Great Dodd gather, and when that gather was off, word-of-mouth meant we were topping the lists for getting hired for other farm's gathers too.

Finally, we had earned respect, and if this journey had only ever meant achieving that, then that would've been the moment I knew I was done. But it wasn't. I had long arrived at a place where sheep farming meant more to me than anything else I had ever worked to accomplish, and I wanted to keep pushing myself to get even better. There were always going to be other dogs to train. Mind you, I did allow us to rest on our laurels for a moment. I was so proud of my dog.

Regardless of how I actually felt about local respect and recognition, it turned out that it wasn't just in my little patch of Cumbria that our work was getting noticed.

Chapter Fourteen
SAS: Ewe Dares Wins

I could not believe what I was hearing when my mobile phone buzzed in my pocket out in the field that day. 'Hannah Jackson? We are thrilled to tell you that you are on the top-ten shortlist for the 2018 *Farmers Weekly* Young Farmer of the Year Award.'

It just didn't seem real. I'd been watching those awards ever since I had started in farming, and had celebrated the triumphs of friends and colleagues from afar, but never once did I consider I might actually make the list one day. Apparently, I'd been nominated by one of my followers on social media, for both my unlikely journey into the profession, and the work I had done to raise the profile of sheep farming – and women in farming – through my social media accounts. Being nominated was a real honour just in itself, but to get shortlisted by the actual judges felt like pure fantasy.

I was incredibly and undeniably proud, but I certainly thought the credit of making the shortlist was going to be where it all ended. A trio of judges came to interview me, and then I heard the mind-blowing news that I had made the top three. A BBC crew came to Brookside to film me for their *Farmers' Country Showdown* series, alongside all the other finalists: a 28-year-old arable farmer called Matt; and Jacob, a close friend who worked down in the Welsh Valleys, running a mixed farm with his family.

A trip to the iconic five-star Grosvenor Hotel in London beckoned, with a glitzy award ceremony attended by the top brass in British farming, to find out just who had won the coveted award. It should go without saying that it was absolutely unmissable; yet a month later, there I was, stood in Heathrow Departures, about to check in to an international flight for an as-yet-unknown destination, and ringing the people running the *Farmers Weekly* awards to tell them that I wasn't going to be able to attend after all. I'd chosen instead to get broken apart by the British Special Forces on national television.

Channel 4's *SAS: Who Dares Wins* first hit television screens in 2015 and was an instant sensation. The show's format is fairly straightforward: take 25 recruits from the general public and put them through a trial of tasks designed to replicate aspects of the selection process for the British military's most notorious Special Forces unit, the Special Air Service (SAS). The challenges intensify the deeper the contest goes, both physically (as the toll of sleep deprivation, fatigue and hunger begins to eat away at the contestants) and psychologically (as the gruelling tasks all take place in an environment designed to replicate a real-life conflict situation).

Inevitably, the number of recruits gets whittled down as the days pass. You are either thrown out by the instructors – all ex-Special Forces soldiers themselves, led by Ant Middleton, a grizzled and heavily tattooed hardman, who has zero hesitation in dismissing anyone that he deems to be unfit or unworthy – or you can 'VW', Voluntarily Withdraw, at any time by handing in your unique selection number worn on your sleeve.

The series climaxes at the end of ten absolutely brutal days with the forced 'capture' of the final recruits standing. They have their arms bound and faces hooded, are subjected to white noise and stress positions, and they are tested, repeatedly, for their resistance to interrogation. At the end, whoever is left is told whether they have passed or failed; and that's pretty much it.

For most normal people it sounds like their idea of absolute hell on earth, but the more series the show racked up, the more I couldn't help but be drawn to the idea of this 'ultimate' test. I was fascinated by what people could push themselves through. The selection process, albeit a watered down and highly compressed version of what the SAS recruits actually have to endure, was as much an examination of your mental make-up as it was your brawn. It wasn't just about passing or failing, either, it was about truly finding where your breaking point lay. You could see that, for many of the contestants, finding out that point lay way beyond what they had previously thought they were capable of was life-changing.

My job in farming is undeniably physical. There are days when the combination of hard tasks in awful weather can really push you to dig deep within yourself; but as tough as it could be, I hadn't met my own physical breaking point, and didn't feel I had ever really come close. It was not something I could ever afford to test. I had animals that were absolutely reliant on me being able to do my job, but *SAS: Who Dares Wins* gave participants the opportunity to safely find where their limit was, and I had always really wanted to apply.

There was only one problem: women weren't allowed. The SAS had been an all-male regiment throughout its history, which has its origins right back in the Second World War. Women were not seen as being physically capable of getting through the six punishing months of selection, which included the infamous 'Endurance' test, in which candidates must march 40 miles across the Welsh mountains, with full kit weighing 25 kilos, in 20 hours or less. Many ex-SAS members had gone on record to say that having women in the force would damage the established, and hitherto successful, all-male team dynamic. Under 10 per cent of the male applicants were successful anyway, and that's in spite of them having glittering service records in regular military regiments, so what realistic chance did a woman really have, they argued . . . incorrectly, I felt.

In 2018 everything changed. Mounting pressure to extend equal rights to female members of the military saw a change in the rules. Women were finally allowed to apply to the SAS for the first time and, in a nod to this monumental historical event, *SAS: Who Dares Wins* opened up its applications for women to compete alongside men too.

I had already been nominated for Young Farmer of the Year when I first submitted my application to *SAS: Who Dares Wins* but, in a display of a complete lack of military attention to detail, I accidentally sent it to the email address requesting formal removal from the application process.

My heart sank to my boots when I realised my mistake, especially as the application deadline had already passed. What an absolute Grade-A first-class tit. Not the best advertisement for 'Girl Power', was it? Immediately, I sent the most grovelling apology of my life to the production company, and rang their offices for good measure. There is an obvious lesson here about always reading the instructions on an application for anything, but I do think that I ultimately underlined to myself, and them, just how badly I wanted it. There's a fine line in life between being self-confident and a massive big-head, but I just knew I could do it, and didn't mind telling them explicitly that I genu-inely thought I was the person they were looking for.

Sometimes in life you have to put aside any very British self-deprecation sensibilities and just go for it. I had nothing to lose by that point anyway. Well, nothing apart from a luxurious night in the Grosvenor Hotel, a beautiful meal cooked by top chefs, a glamorous awards ceremony, and the shot at potentially being crowned Young Farmer of the Year.

It was so cold outside that it hurt your lungs to breathe, and breathing wasn't easy anyway as we were thousands of metres up in the sky. High altitude and sub-zero cold, I would soon learn, bleeds energy from your body at an alarming rate. This was not Cumbria, and I was not Hannah Jackson. I was num-ber 25, of the 25 men and women brave (or foolish) enough to sign up for this.

Massive mountains loomed over me in every direction I looked. This was the Andes, after all, the longest continental mountain range in the world, stretching over 4,000 miles from top to bottom (about seven times the length of Britain), a remorseless expanse that spanned most of South America.

The bright yellow bus I was sat on with all the other 'civil-ians' was carefully picking its way up a rock-filled gorge, climbing ever higher into a whitewashed wilderness. The few mountain faces that weren't covered in thick snow felt dark and forbidding. Frigid frozen waterfalls hung down over us like the Sword of Damocles. I blew hot air into my hands for warmth, and tried to remember the reason why I was here at all.

The last few weeks back home had passed in a blur. I'd been given a 'last-chance saloon' Skype interview by the production company that makes *SAS: Who Dares Wins*, which I passed, followed by a gruelling fitness test in Halifax, a psychological examination in London, and a medical, before I finally received the confirmation that I was in, by the skin of my teeth. I was posted two pairs of military-style boots to break in, and had just enough time to get my vaccinations, but it wasn't until I was at Heathrow Airport, and was handed back my passport with a freshly minted Chilean visa stuck to its pages, that I discovered where we were actually being sent. It was just the beginning of the process of keeping us all in the dark about what we would be doing and when things were going to be happening. In real wartime, there are no warnings. I had just enough time to hurriedly text my mum – 'I'm going to Chile' – before myself, and all the other recruits, were swept up onto the plane.

The day after arrival in Santiago, the Chilean capital, we'd had an off-camera acclimatisation run up a mountain. It hadn't gone brilliantly. I had been about average but some people had really struggled, so we were told by the production team we were all going to need to do at least another two hill runs before the show's filming could even begin. The next morning, we had scooped up all our gear, boarded that yellow bus, and were driven for several hours into the mountains.

I'll tell all of you right now: what you might see on your screens when you watch *SAS: Who Dares Wins* is just the tip of the iceberg. At no point does the mask slip. There is no TV fakery or tricks. The immersion is 24 hours a day, seven days a week, regardless of whether they are filming you or not. It is a non-stop commitment from start to finish. You never know when or what is going to happen, or where you are heading next.

It was brilliantly done, to the point that pretty much from day one you feel so absolutely consumed by the entire experience, you are totally convinced the 'wartime' recruit training scenario is completely for real.

I leaned on the steel frame of the seat in front of me and chatted to another of the new recruits. Esmee Gummer, number 23, a Londoner with fluffy blonde hair and kind eyes, soon became my closest friend through the process. I can't remember what we talked about; home and silly stuff, probably, just killing time and nerves. The next thing, the brakes of the bus are slamming on, there's a huge flame-filled explosion to the side of the road, and men in balaclavas are leaping on the bus, screaming, 'Get your fucking heads down, right now!' I did exactly as I was told and, as absolute pandemonium broke out all around, I felt a gloved hand close around the neckline of my coat. *Well, it's happening,* I thought to myself, while staring down at my crotch. *I guess we're not doing those training runs, then.*

I'd made so many plans going into this. One of the first things you notice as a long-time watcher of the series is that a recurrent trait shared among the most successful contestants is that they inhabit the character known in military parlance as 'The Grey Man'. The Grey Man is a person who is forgettable, seemingly inconsequential. They don't stand out by finishing first or last. They are unlikely to receive praise or punishment, or to get rigorously interrogated, as they appear utterly unimportant. Generally, it means they progress, they survive, and they pass. Taking the transition to Grey Man very seriously indeed, I'd let my red hair fade in the weeks before I left, and I even told Mum that if she was called for interview at any stage, *not* to give them any information that might hurt me or my chances.

I knew she was taking it seriously too; she didn't even dare risk giving me her traditional handwritten letter of love before I went away, just in case it could have been used against me at a later date. My strategy was utterly set on becoming the Grey Man.

My coat's hood was pulled over my head and I was led off with my hands held together but, unfortunately, as I stumbled on the final step out of the bus, I let out a little nervous smile. 'You fucking think this is fucking funny, redhead?' screamed one of the abductors, directly into my face. 'Fucking kneel down now!' I was thrown into a line with all the other contestants. *Nice one, Hannah, you total dickhead*, I thought to myself, now looking down at the compacted snow between my knees. My entire 'Grey Man' tactic was blown out of the water within 30 seconds of the start.

'Welcome to Hell.'

I knew that voice right away. It was Ant Middleton, the chief instructor, former Afghanistan and Northern Ireland veteran, and a member of the Special Boat Service (SBS). We were told to look up, and there they were. The notorious 'Directing Staff', known as the 'DS'. People that I'd only ever seen before on the television up to this point. Stood next to Ant was Jason 'Foxy' Fox, former Royal Marine Commando and sergeant in the Special Forces, and Mark 'Billy' Billingham, who was in the Parachute Regiment and a Mountain Troop specialist in the SAS. They had been the ones pulling us off the bus, but one person was missing from the line-up: Ollie Ollerton, Royal Marine and former SBS Special Forces frogman.

'Stand up,' commanded Ant, sternly. 'Follow the DS.' He pointed out behind us, and there was Ollie, halfway up an Andean hill, turning to run away from us. Everyone took a beat to process what the hell was happening, but it was a beat too long: 'Go on! Move it! Move your fucking selves! NOW!' Ant, Foxy and Billy started to physically push us from the spot and everyone started running.

The feeling was intensely surreal, but there was no time to get starstruck or even slightly self-conscious at being filmed: we were off, it had begun, and the only way out now was if I VW'd and gave up my number. I decided there and then that, no matter what happened, there was absolutely no way I was ever going to let that happen. I might have made a mess of my Grey Man routine, but they'd have to physically tear my number 25 from me if they wanted me out of the process and off this show.

Going into *SAS: Who Dares Wins* I felt there was a real point to proving that women actually could do it. I cared far less about being worthy of the SAS than I did about demonstrating that women were strong, capable and psychologically on point. I might have gone some way to proving that in my little corner of Cumbria, but the women with me on this show now had the chance to underline it to an audience of millions back home. What I realised pretty quickly, though, was that there was already a basic degree of equality cut into the unwritten code of the military. When it comes to training, you are all equal; because you are all worth less than dog shit.

Everything on *SAS: Who Dares Wins* was designed to remind you that no one is better than anyone else. Your names are taken on day one and replaced with your numbers. Everyone is dressed the same: grey hoodies, combat trousers, boots and a black beanie hat. And everyone only gets two sets of clothes: one wet, one dry.

Everyone carried identical kit that weighed exactly the same:

25 kilos in one military-style bag called a 'bergen'. We ate the same porridge-like cement for breakfast, we slept on the same crappy military canvas beds, we washed the same (no hot water, no shower), and we went through the exact same training exercises together. Even the basecamp itself was nondescript. It was a former military camp. Goodness knows how old it was, but it was made up of a series of crumbling concrete bunkers 2,500 metres above sea level, with a parade ring simply marked out by white stones. The whole thing blended into the slate-grey, snow-white Chilean landscape perfectly; just like we were supposed to as well.

If you didn't like any of it, you were out. The first recruit to get the axe didn't want to go to the same toilet as women for religious reasons and, pretty soon, it was abundantly clear that the only way forward was to try and gently morph into the same set of stinky, shitty people. Forget any sense of individual self-worth or beauty – I swear I started to go hairy after about a week, and still shudder at the tragic results of trying to brush my hair with a shoe-polish brush and tub of Vaseline – our success as individuals on the show relied on first being successful as a team.

One of the simple ways they grind you down early on is by 'beasting' you. It is a routine as old as the armed forces themselves, and is overtly a way for your superiors to find ways to break you, humble you, or generally just test your mettle, by forcing you to do arduous training exercises under stress and duress. Subversively, though, it is another means of bonding you to your team through the totality of a shared experience or, to be more accurate, shared suffering.

Beastings would always happen if one person had done something wrong: failed on an exercise or not been attentive enough (according to the DS), but they would also happen regularly for no obvious reasons whatsoever. Often, we'd be lying in the middle of the night, trying to sleep in spite of the cold, and the bright lights of the bunker would suddenly burst on. The DS would then pile in, screaming at us to get up, get outside and start performing burpees, doing body-weight exercises or just running on the spot, in the pitch-black, with freezing digits and exhausted limbs; and always with the nagging background call of 'Remember, you can hand in your number and VW at any point.'

Another method of achieving the same results was the all-night-long 'stag duty' we were tasked with. It saw us all placed on a strict rota, where we would have to spend one hour outside keeping watch for the 'enemy'. The shifts were designed to overlap, so you'd spend half an hour with one person, then half an hour with another, before you could come inside and wake your replacement. It meant the bunker door was opening and closing into the ice-cold Chilean night every 30 minutes and, given I was number 25, the last to be recruited and the last in the line of beds, I found I was placed right next to the swinging door. I lay awake night after night swearing I would never ever mess up an application process so badly again in my life.

This idea of 'military-style equality' didn't sink in straight away, but we'd all taken a huge step in the right direction by the end of the first big task.

We were pulled out into the parade ring in full kit with our 25kg bergens on our backs, and told to get running. The run took us high up above the basecamp and well into the highest mountains of my life. Honestly, had I not been so knackered and anxious, the vistas up there in the Andes would have registered as some of the most stunningly beautiful I'm probably ever likely to witness. We ran downstream alongside this spectacular white-water river that hurtled into blackened boulders, before we were eventually brought to a standstill at the point that the river transformed into this simply epic waterfall. It erupted from the hill like a broken faucet, pouring foam and water into a deep toilet-bowl-shaped plunge pool that had been hollowed out like a dead man's chest.

'If you fall in there, you are going to die,' said Ant, by simple way of introduction. 'If you can't get out, there's no point you continuing on this course.' Lesson one: pretty much every exercise on *SAS: Who Dares Wins* was counterintuitive to anything you would ever do in ordinary life. Everyone knew, despite the now obvious dangers, we were all going in.

I watched the first woman, Nadine the firefighter, pull off the exercise well; but she was followed by Saranya, a 19-year-old student who wanted to become a Marine, who panicked almost as soon as she hit the cold water. She received no sympathy, just a mild bollocking from the DS, and then it was my turn. They tied a red climbing rope around my waist and in I went, directly into the waterfall's plunge pool. The water was made up of snow and glaciers melted directly off the mountain slopes – basically, as close to freezing as you can get without actually being solid – but there was no time to think about anything other than completing the exercise. We were told that hypothermia could set in within minutes, so there was more than just a competitive reason to get the whole thing done fast.

I submerged myself on instruction and emerged with only my head above the water level. 'What's your name! How old are you! What's your number!' My brain might have been gently freezing, but my mouth was working well enough, as Ant Middleton clung to the dry end of the rope and barked out at me like some deranged Arctic ice-fisherman. I answered the final question about my sister's names and climbed out, feeling a bit pleased with myself. It was cold, but it was cold for everyone, and I hadn't flailed my arms about, panicked or wasted energy.

We were then sent to a warm tent to get changed immediately. 'Get out of your wet kit and into your dry kit! Now!' We all, obviously, happily obliged. The next thing I knew, Ant Middleton and the rest of the DS came storming around the corner and everyone was made to line up outside. They looked absolutely thunderous. It was blatantly obvious we had made some massive blunder, but no one knew what it was. We stood in silence, awaiting the rollicking.

'You, step out. You, step out. You, step out.' Ant made his way along the line, picking out nearly all the women from the men, including me.

'Look at the state of you.' I looked down. The wet bra underneath my one set of dry clothes had leaked water right through my grey T-shirt. 'You have been given a dry, warm tent,' he began, 'A DRY FUCKING WARM TENT TO GET READY IN AND YOU COME OUT LIKE THIS!' He was snarling, spitting, inches from our faces, blue with rage. 'YOU'VE JUST COMPROMISED YOUR WHOLE FUCK-ING TEAM!'

It was made absolutely clear to us all that having a bra or breasts was no excuse for getting your dry kit wet. We should have taken our bras off, squeezed them out or put them in our bags, anything but put them back on and get our only dry set of clothing wet. Once your dry kit is wet, especially in a cold climate like that, it isn't getting dry again easily and pretty quickly your whole body will get as cold as if it were back in the water itself.

It was a massive mistake, and the whole group, men and women, were to be beasted because of our error. Ant ordered everyone to hold their 25kg bergens straight-armed above their heads in a weight-lifter's high-press position. 'I didn't want to single you women out. To have a divide,' said Ant, in an oddly conciliatory tone. 'But you fuckers are making me do it.' That was more like it.

Day one. The 'induction' was done, and now we all knew our place.

As the days passed, I came to discover that my life as a sheep farmer had given me a real advantage. I had 'hill legs', the ability to run up slopes and find a rhythm, from all my days running around the high fells after stray sheep. When it came to lifting exercises (the bra incident certainly wasn't the last time they had us holding our bags aloft, but sometimes it was rocks or people, just to mix things up), I could always find ways to power up from a good strong base, from all the time spent hauling sheep about in pens. I was a strong swimmer naturally, but I definitely fared better in the cold generally, thanks to my work outside in the harsh Cumbrian winters; and sleep deprivation over multiple days wasn't anything like as big a problem for me as it was for others, due to all the zombie-fied training days I had endured during lambing time.

Sheep work, I recognised, had embedded a deep strength in my muscles and over in Chile, I could see from the other contestants, especially the ones who had exclusively built their bodies indoors in gyms, that I had a lot to be grateful to farming for.

Psychologically, one of the biggest adjustments in the first week of *SAS: Who Dares Wins* was just getting used to having someone constantly shouting at you, and never giving you a scrap of praise. Okay, the culture of sheep farming isn't quite *that* brutal, but clearly, I was one of the only contestants that received zero overt praise in their regular job anyway, and I already knew it was nothing to take personally. In a curious sort of way, I actually discovered there *was* a subtle, but definite, cross-over with the military and sheep farming. There was a sleight-of-hand that came with a heavily subverted compliment, if you only knew how to identify it when it was there. One day I remember being out on one of the big runs and finding myself slightly lagging after a huge summit dash. Ant Middleton ran up next to me and screamed: 'Fucking hurry up, number 25! You don't want to be like one of the shit c*nts at the back!'

There it is. That's the compliment, I thought. *Number 25: Not a shit c*nt.*

In the first four days four people voluntarily withdrew, and as the first week progressed, I became aware that more people were dropping off all around us. I was never truly in a moment where I felt wholly comfortable with my place – the show does that, it constantly makes you feel like you could be taken out at any time – but I felt secure enough to know that I wasn't going to humiliate myself, and even allowed myself to enjoy a few of the tasks. Tasks I knew I would never get to do in normal everyday life.

I abseiled down a cliff face-first, running as fast as I dared. I walked over a ladder suspended over a precipice without pausing for a beat. I jumped off a high bridge on command. I swam under the water beneath a giant glacial lump of ice, and I successfully completed an 'ice axe arrest', where we had to deliberately free-slide down a snowy slope and halt ourselves with only our ice-pick.

The more I gave myself over to the process, and to the members of the DS, the more bearable it became. Slowly, I started to tick off the days.

The short, sharp, high-adrenaline exercises always went well. I found I was able to hone in on the simple instructions I was given, and just execute the task without thinking, or worrying, too much about it. Those types of task, alongside the basic beastings, tended to dominate the early days. It was all about cutting through the people who were just never destined to make it in the process as quickly as possible, but, as the final few days approached, and the recruits had almost halved in number, different tactics were employed by the DS, and things soon started to get a hell of a lot rougher.

Chapter Fifteen
SAS: Mind Games

One of my weaknesses was exposed on the longer solo endurance exercises. A classic example came when the DS would send us out for long runs alone; way out into the oxygen-light-air and uniformly lunar landscape, with no obvious finish line and no one around to keep you going.

On previous shows I had watched them actually make the recruits run until someone dropped out or VW'd, so that was always in the back of my mind, but in the front was this constant nagging voice telling me that I was struggling, that it was too hard, that I wasn't going to make it, that I should think about giving up.

By this stage in the contest, everyone was clearly fit enough to get through, so it was those negative voices in your head that the DS looked to exploit and turn up, to get more recruits to quit. As we headed towards the climax of a challenge that I knew could only last a maximum of 11 days in total, the mind games seriously took over.

It got to the point where I knew I had to find a strategy. I had to trick my brain into being able to fight with me, to suppress that poisonous negativity, to really start believing I could do it: no matter what. *If* I got through these next few days then I knew the biggest mental test of my life was coming up: the infamous 'Interrogation'. And if I carried on like this I was going to stand absolutely no chance of getting through it.

Have you ever seen the Disney film *Mulan*? Don't laugh. Okay, in *Mulan*, the Hun army have crossed the Great Wall of China and are marching down to kill the Emperor and put an end to his dynasty. In rural China lives Mulan, the only child and daughter of her war-hero father. Her father is old and injured, so when forced conscription comes, stating 'one man from every family' must join in to fight the Huns, Mulan steals his conscription papers, cuts her hair, deepens her voice, and signs up to military training in his stead. It is against the law for women to fight, and the penalty for deception is death, so Mulan realises during her basic training that she needs to get stronger and fitter to maintain the ruse. She really struggles at first and, just as she is on the verge of being dismissed altogether, a training montage begins, underpinned by the song 'I'll Make a Man Out of You'.

Fine. It's undeniably cheesy, but the message of the film, and that moment, was not 'To win, you need to become a man', but that the strength we need to achieve all our goals lies within every single one of us, no matter what your gender or background. If that wasn't a symbol of *exactly* what I was going through on *SAS: Who Dares Wins* then I'm not a twenty-something sheep farmer from the Wirral. The song progresses, Mulan starts kicking butt, and by the end, she's the strongest recruit in the entire platoon.

So, on those long gruelling solo mountain runs I took to singing 'I'll Make a Man Out of You' to myself, and by the time we were called back to basecamp, I *was* Mulan. Fully capable of getting through *SAS: Who Dares Wins* and single-handedly saving the Emperor and the entire Chinese nation.

'Hannah?' I turned on my camp bed. It was my friend Esmee,

number 23. 'Sorry to ask, but are you singing *Mulan*? Out there, on the runs?'

I had been exposed. 'Erm, yeah.' There wasn't any point denying it, especially not to my best mate. 'I just feel it helps me be a bit more determined. You know, that I should be more like Mulan, and less like Hannah Jackson.'

Esmee smiled back at me. 'We must be swift as the coursing river,' she started to sing. We both burst out laughing. That was it. The *Mulan* sisters. From that moment forward, whenever we felt a bit low, or tired, we would sing a bit of *Mulan* and find that inner steel.

Whole careers have been made off selling books and speaking tours with long, deep, meaningful metaphors for life, but all you really need in the toughest moments are good friends and a bit of Disney.

Mulan might have helped me solve that little mental Rubik's Cube, but it was no help whatsoever when it came to the one task I feared above all others.

'Milling' is a military training exercise where you square off against another recruit, of a similar size and weight, and fight each other as hard as you possibly can. You can only punch, preferably in the head, and you are not allowed to cover up, back off, evade strikes or stop throwing your fists until the DS has told you to stop.

It teaches unbridled aggression and is a test of both your fighting spirit and courage, with the rationale that, in a war, soldiers who choose to take cover rather than fight back are a detriment and a danger to their entire squadron.

This was the first time I felt one of my other major weaknesses was about to get laid bare. I am just not an aggressive person. In fact, I hate fighting. I was never going to join the military as I could never kill someone. I was competing on the show to test my physical and mental limits, and not my capacity to commit a violent act.

I had a lot of respect for the military going in, and having just a taste of what it took to be an elite-level soldier had elevated that feeling massively. Being in an actual war, though, was almost incomprehensibly hard for me to contemplate and, regardless of any opinion on the validity of some of the fights this country has been involved in over the years, I couldn't help but look at all our military veterans with anything other than real reverence after my experience.

I knew, before I had even applied, that I had the luxury of not being forced to actually fight someone in a war, but I also knew that the milling exercise would come around eventually. It had been in all the previous series, and this was going to be no different.

'You're no longer a team. You're no longer friends,' began Ant Middleton, as we all stood in a row, with the makeshift boxing ring, its corners marked with four stones, laid out on the dirt in front of us. 'That person in front of you is the enemy.'

I stood and watched. The anxiety and fear grew inside me with every knock-down, bloodied nose, victory and failure. It wasn't that I was afraid of getting hurt either, I just couldn't put aside my friendships with these people and hit them in the face. I was terrified of causing someone I liked real physical pain, but I knew if I didn't take part then I would definitely be thrown out. I was trapped and, for the first time in the whole process, I felt my bottom lip start to quiver.

'The enemy doesn't care what fucking gender you are. What religion you are. What race you are. They just want to fucking kill you. Full stop.' Ant was back on the prowl. Number 21, a female recruit, deliberately chose number 16, a man, and was beaten until her mouth and nose were blackened. It was horrific, but it was this 'military equality' they were trying to drill into us. No one was to be given any special treatment. Number 16 felt absolutely awful afterwards. He was almost on the verge of tears. 'This is a head-fuck, man,' he said. He wasn't wrong.

'Number 23,' shouted Ant, and up stepped Esmee, looking as white as a sheet. She started staring up the row at who was left. I could see she didn't know what to do, but in this absolute no-win situation, I felt I did. She caught my eyes and I lifted my eyebrows to her. 'Pick me,' I tried to motion, without the DS noticing – so she did.

I thought Esmee might be thinking along the same lines as me and, given the DS had never seen either of us in a fist-fight, I felt it was our best shot at pulling off a bit of a hustle. There had been points already where we had really needed each other, but this was our biggest test. I just wasn't going to hit her that hard, and I knew she'd never grass me up or make me look foolish as an opponent: it just wasn't in her make-up. She was hard, though, in fact I knew she taught boxing as part of her work as a physical training instructor, but at least it would only be me getting hurt if she hadn't quite got the message.

Foxy put on my headguard and gloves. I looked at Esmee and nodded reassuringly. Then I looked back at Foxy, and found the words 'I can't do it' spilling right out of my mouth. 'Yes, you can,' he fired back sternly. I started breathing hard, almost hyperventilating. *Shit, Hannah,* I thought, *you're actually going out here.*

'Open your mouth and put this mouthguard in,' demanded

Foxy, then, in a rare break from the absolutely uncompromising, no positivity, DS party-line, he leaned in and simply said: 'You'll be fine.'

It was fine. We got through it with a couple of bruises, but neither of us truly let rip on each other at any point. I got back to basecamp and began to cry. That evening I was called in to face the DS for a debrief. It was something that had happened regularly with the other recruits, but this was my first visit. A sack was put over my head and I was taken to a back room.

'What happened today?' It was a fairly straightforward question that revealed they already knew I'd tried to pull the wool over their eyes, but I tried to bluff my way through anyway. Foxy stopped me mid-flow: 'You didn't put in your 100 per cent effort with the milling, number 25.'

I told them the truth. I had to. I said we had spent so much time bonding as a team. That we were constantly being told by the DS that these were the people we had to look out for. To protect at all times. Then we suddenly have to put all that to one side and batter each other, as if none of that really mattered at all.

'But this is part of the process,' they replied pointedly. I paused for thought. I was walking a very precarious tightrope that could lead to my expulsion from the course.

'But it isn't what you do in real life, though. Even at war. Is it? Not to your friends,' I offered.

They stared back at me impassively. Absolutely inscrutable poker faces. Every one of them.

'Guard!' someone shouted. The bag was placed back over my head and I was led back to basecamp.

For me, this wasn't about warfare, it was about proper

friendship. Luckily, I was allowed to continue, with that one blemish on my record.

As the end of the process grew near, I lost track of time. By this point we were hanging on grimly, and I knew I didn't have much left in the tank. The final eight people, four men and four women, had already proven they had what it took physically and mentally to progress. Now it was more about holding out for Interrogation and praying you didn't get badly injured.

SAS: Who Dares Wins always finished the same way. The final task began at some point on day nine with 'Escape and Evasion'. The remaining recruits would be sent into the wilderness and pursued by an enemy. At some earlier date, they would have already been given two stories to remember: one was a red herring, a plausible non-threatening excuse for being in the 'warzone'; the other was the real reason you were really there, usually a variation on being in a military regiment tasked with attacking and eliminating the 'enemy'.

Capture by the enemy was inevitable, and followed immediately by being violently bound and hooded. The Interrogation stage would usually be under way by day ten. Already you would not have slept or eaten for at least a night, thanks to Escape and Evasion, and Interrogation itself usually lasted for at least another 18 hours. You would be hooded for almost the entire time you weren't being interrogated, so you could not see, and you would not be given any food, nor could you sleep. You would also be placed in stress positions while white noise would be played at ear-splitting levels between the interrogations, meaning you couldn't hear, or barely think. But all of that was actually marginally more comfortable (or just less awful) than the interrogations themselves.

When it came to the interrogation a mixture of questioning techniques including anything from screaming a barrage of questions at you to a (probably more disturbing) 'good cop' routine were employed. The intention was always to either extract the truthful information from you, or to break up the team, or just to break you psychologically as an individual. At any stage you could VW by calling for a secret man, known as 'The Umpire', who acted as the arbitrator; but even if you did make it to the end of the 18 hours, if the DS didn't think you had performed well enough, you could still be handed a 'fail'.

The exercise was always intended as the ultimate test of everything we, as recruits, had been through up to that point. Your ability to 'resist' was the final piece of judgement on your ability to be in the SAS required by the DS. It wasn't as straightforward as saying nothing, though. Complete non compliance in an actual capture scenario would quickly lead to torture and death. This was a game of finding out who could say the absolute bare minimum, who could be compliant to a point, without ever turning in their friends, who could stay alive, who could stay sane.

I would never have believed I would be praying for that ordeal to begin, but I had badly hurt my knee during a fall on what turned out to be the penultimate big hill run, and as insane as it might sound, I was actually looking forward to a bit of a sit-down.

My heart sank when we had then been led to the bottom of a hill for a final timed 5km run with full kit. It was explained that if we missed out on the expected finish time then we were out. I'd strapped up my knee and was digging as deep as I thought I possibly could, but I just fell further and further behind. 'Come on, number 25, dig in!' shouted Foxy, who appeared next to me. I was at the back by this point, and Esmee was right there with me too. It looked like time was about to run out on us both.

I knew this was it. That I had finally met the point that I'd gotten into this whole bloody process to discover. What was I going to do now? Slow down and give in, or just try, one more time, to dig even deeper than I ever thought was possible? I started to pull away from Esmee. I pushed through the pain barrier in my knee and found a way to just move forward. I couldn't face failing without knowing I had given absolutely everything I had, even if that meant passing out or collapsing on the trail. I forgot about any other task, past or present, gritted my teeth and just ploughed forward relentlessly until I eventually summitted the hill.

Esmee came in just behind. 'Number 23,' called the DS. 'Give me your number. You have been withdrawn from the process.' I had made it, just, but my best friend had failed.

I stumbled into Esmee's arms, choking back the emotions. 'I'm really going to miss you,' I said, quietly. She pulled me in and whispered in my ear: 'You will make it to the end, Hannah. Do. Not. Stop.' Then she was ordered off the hill and was gone for good.

Escape and Evasion started up soon enough.

We were given one piece of fruit each for dinner on day nine, then, shortly after dark, chucked onto a truck and driven out into this deep-bottomed valley for a night exercise. We were given a small map and split into two teams of four to navigate to a rendezvous point, but it quickly transpired that we were actually being pursued by members of the Chilean military with attack dogs on leashes. It wasn't until we spotted their flashlights behind us that it dawned on us all that Escape and Evasion, the end game, was very much under way.

My team were the first to make it to the rendezvous point, only to then be told that the enemy were closing in on us, and we now had to run to a second rendezvous point about a mile away. We were, gloriously, given a bread roll to share between us and sent on our way with a stirring 'Hurry the fuck up!'

Halfway to the second rendezvous, the Chilean military pounced on us from all directions. Flashlights, dogs and powerful hands forced us to the floor, as our own hands were bound with cable ties and sacks were placed firmly over our heads. At roughly the same time the other team had also been captured. Escape and Evasion was the only task we were set that was impossible to pass and, as my head was hooded into darkness, I felt the same way as I had when the initial explosion had gone off right back on day one: *It was happening*. I had made it to the most feared of all the training exercises.

Interrogation was run by the DS and a new, highly specialist team, steeped in the art of this particular SAS selection exercise. We were immediately placed in stress positions: hands still bound, hoods on always, unless we were being interrogated. I was placed in a seated position with my arms outstretched in front of me. Naively, I thought it wasn't too bad, but pretty soon the lactic acid started to build up and I was in absolute agony. I dropped my arms for a moment, and an unseen hand was on me instantly, forcing me back into the position.

In the first stage of formal interrogation we were individually stripped of all our kit and clothes. The cold and vulnerability of being stood only in your underwear was a pretty crude, but very effective, attempt at getting you to give up right away. One by one, we were questioned and insulted by a new crew of rough-looking men, before we were thrown a prisoner-style boilersuit to put on. After that we were blindfolded with the hoods once more and placed back in the stress positions. This time we had headphones placed on our heads, too – drilling noises, crying babies, squealing pigs – funnelled right into our ears for seemingly endless hours.

Unable to speak to each other, and with no knowledge of what anyone else had revealed, or not, we were kept in the literal darkness, our limbs constantly readjusted into new stress positions by unseen people, or placed back every time we broke from our position in pain.

I knew beyond any doubt that there was no way they would ever get any information out of me. Either the real 'story' of what we were doing – that we were British military looking for a weapons cache – or the cover story, that we were in Chile searching for three British tourists who had gone missing in the mountains, as part of a team sent over by Cumbrian Mountain Rescue. The reason I knew I wasn't going to be giving anything away was not because of my incredible strength of mind or supernatural abilities at resisting interrogation. It was because I had almost completely forgotten both stories, some time ago. Sleep deprivation, physical exhaustion, and the fact I could not confer with my teammates, meant the chances of me retrieving any semblance of the facts, or the fiction, from my head was absolute flat zero.

The one thing I had managed to recall was the bit about being in Cumbrian Mountain Rescue. Hardly Einstein-levels of intelligence given it was where I actually lived. I had something of a plausible justification, though, and told the interrogators that I had been recommended to come to Chile by a local Cumbrian farmer.

'Who was that exactly, then?' they barked.

'James Rebanks,' I replied, panicking, and, in proof of how far gone I really was by that point, I immediately started to worry about the safety of my great friend James and his family, who I was pretty convinced were now about to be hunted down and executed by the enemy, because of me.

The whole final test was designed to see if, when you are left alone in charge of your own mind, you could find ways to control it. Well, I already knew I was going to fail due to not being able to give any useful information whatsoever – a deadly strike-out in any actual interrogation scenario – so I decided to just set about hanging on for as long as I possibly could. My memory might have let me down, but I wasn't going to let myself down further by VW'ing. That was the one thing I had decided right at the start, 11 days ago now, and I had zero intention of breaking that promise, having come so far.

Six hours in, and as day broke on the next day, three of the remaining eight recruits had already VW'd, so we were moved on to the next stage of formal interrogation with just five of us left: two women, and three men.

This time the interrogator was female, short in stature, with an even, headmistress-like tone, but armed with deep-blue eyes that I felt were piercing my soul. I could remember the DS Ollie Ollerton saying something about the art of being able to convince a trained interrogator that your cover story, your lie, was the truth, and that a good prisoner should spin out the lie for as long as humanly possible, before they are eventually forced to tell the truth. Well, given I could barely remember the cover story, let alone the truth, I was on a hiding to nothing with this woman. I might have been at the worst end of the scale with my memory fail, but the truth that actually transpired was that not one of the final five had given anything close to the same account of why we were actually in the country.

Following another round of blindfolded stress positions, I was hauled into another room. I had my hood removed and was very surprised to find myself stood facing my four (still-hooded) friends, for the first time since we had been captured. The interrogator this time was a man with neat grey hair and a goatee beard.

'I'm leaving this room now,' he barked. 'You're in charge. Wait here.'

Stood in the room, the only one of my team able to actually see, I noticed that in one corner there was a table with water and cereal bars placed on it.

This was the brutal brilliance of this final act in *SAS: Who Dares Wins*. You just have absolutely no idea what you are supposed to do. If I gave my friends food, had I just signed my own death warrant for taking things without explicit instruction from the interrogating team? But if I didn't, when everyone had been without food for a whole day and a night, wasn't I just a completely terrible person? My brain was paralysed. I just stood there in the end, totally impotent, doing absolutely nothing.

The interrogator swept back in, looking livid: 'What kind of fucking person are you? Leaving your friends to starve!' he screamed. 'When there's fucking food right in front of you on a table?' The next thing, everyone had their hoods taken off. He asked every single person what they would've done, and every single one responded to say they would've given out the water and food. I felt awful.

We were into the next stage of interrogation: divide and conquer. The interrogator rounded on me again: 'Tell me which one of them you don't trust!' I tried to pull some sort of acceptable answer from the deep fog of my brain. In the end I went with Milo, a quiet and unassuming lad who had wanted to make it into the military. 'Why?' barked the interrogator, with steam coming out of his ears.

'Because he's a vegan,' I replied.

The interrogator burst out laughing. To be fair, I can look back now and properly laugh too. Poor Milo. He was such a good guy. A very thoughtful and conscientious human. I didn't mean it, I just couldn't think of anything else.

We were all sent back to stress positions, but pretty soon after that we were hauled back together again and, this time, everyone's hoods were removed at once. The lady with the penetrating blue eyes was back with the angry goatee-beard interrogator in tow. We were all told our cover story was completely blown, and one of the recruits, Mark, who *was* capable of remembering the truthful details, gave up that we were actually members of the British military.

I certainly did not feel it at the time, but it was, under the circumstances, the correct way to go. Our lack of coherent details meant we could not continue to tell lies – I mean, *I could* obviously, I could have gone on all day without ever revealing the truth – but we had been going for many hours now, and we were getting absolutely nowhere.

This was an important step towards the end. In a real hostage situation, you just couldn't go on telling such obvious lies as a group. One of the final five, Rick, in a later interrogation, said he felt we had all let ourselves down in that moment, by giving away too much. On that basis he was failed for putting his personal pride ahead of that of the team. By that point in the process, the margin between a 'pass' and a 'fail' was as thin as the edge of a razor.

Back in the stress positions, it was almost 48 hours since I had last slept. I was wearing just the boilersuit, in near Arctic conditions, and all I'd eaten was an apple and my quarter of a roll, the previous day. The next interrogation began with me being marched somewhere new. I could feel warmth and, in my addled state, I thought the interrogators had decided to keep me warm with a personal heater. It wasn't until my hood was off and I was being screamed at again that I realised it was actually the late-afternoon sun heating me up.

A new interrogator burst into the room and sent the screamer off to get me a cup of tea. It was a 'good cop' routine. He was big and burly. Northern, with a tribal tattoo on his forearm and tousled grey hair. He offered me some nuts. 'You've got to save yourself here now,' he said, warmly.

On the television, my 'What am I meant to tell you?' response might have made me look like a right hard bastard, but it was frankly of greater use to the production than the more honest: 'I didn't actually learn either of the stories that I was supposed to for this part of the television programme.' I was absolutely ready to completely spill the beans, to tell all, to bare my soul, but I looked inside my head and just got crickets and tumbleweed in return.

That was the point I knew my experience was about to come to an end. He looked me squarely in the eyes and said, 'I'm going to give you one last chance to save yourself.' I blanked again, and that was me done.

*

The next time my mask was removed, the full 18 hours of the Interrogation stage were up, and I was face to face with The Umpire.

'Who am I?' he asked.

'The Umpire,' I replied.

'I am here to inform you that you have been withdrawn from training,' he continued, in a matter-of-fact tone. 'Ant is going to speak to you now.' He stepped aside, and there was a beaming Ant Middleton. He gave me the biggest warmest bear-hug.

'I fucked up a bit, didn't I?' I said, smiling weakly.

'Just a little bit,' he replied, 'but it doesn't matter.' He went on to tell me how well I had done, how I had helped change his opinion on bringing women into the SAS, that I should hold my head up high and remember that, even though I hadn't walked out with the pass, I had completed every single aspect of the course. The rest of the DS also came and spoke to me individually, all underlining the same sentiments, and sending me off on clouds higher than the highest peaks in the Chilean Andes.

Out of Interrogation, and formally out of the process, I had one phone call I was allowed to make right away. 'Oh my goodness, Hannah!' my mum answered, gushing, the enormous sense of relief clearly palpable in her voice. She started telling me how proud she was and then checked herself: 'Listen, Hannah, I can't actually say too much. You're not going to believe this but I'm actually at the *Farmers Weekly* Young Farmer of the Year Awards right now, and the BBC have just put a radio microphone on me.'

I didn't win Young Farmer of the Year. It went to my good friend Jacob from Wales, and I couldn't have been happier for him if I'd have won it for myself, but in that moment, quite honestly, I would've swapped everything I had ever owned or achieved, just for a piece of toast and some sleep.

There was a tap at my shoulder. It was the doctor, leading me off for a psychological debrief.

I'm glad I'm not a soldier. I never could be. I'd demonstrated where I was at with actual violence with the milling exercise. I hadn't entered *SAS: Who Dares Wins* because I was planning on ever joining the military but, in getting to the end, I felt I had helped prove what women were capable of. It was a complete overhaul of my mental make-up, too. Back on the farm, I noticed I had discovered a much greater degree of mental resilience. I have always felt robust as a person, but after the show I felt a lot more level-headed, confident and psychologically strong when it came to dealing with all those things that had unsettled me before at work. Whether it was the online trolls or just handling those times you are presented with pieces of farm machinery that simply do not want to be fixed, I felt a long way removed from the Lazonby meltdown of the previous year, and knew that absolutely nothing in my life was ever likely to be as psychologically challenging as that Interrogation ever again. *SAS: Who Dares Wins* armed me with much more control when it came to facing those harder times in my farming life, but it also gave me the perspective to realise that bad times eventually pass too.

Despite not quite getting the 'pass', I and the other female who also finished, Lou, a doctor and a truly lovely person who did gain the pass, demonstrated that women deserved a seat at *every* table, and not just the elite levels of the military. Women are strong and resilient enough to get through a brutal interrogation and physical selection, but I think we also proved that women have different skills that men might lack. That our ability to see things from a different perspective, to bring a different opinion, to empathise and support, were all things of immense value too.

For the most part, *SAS: Who Dares Wins* provided some of the most consistently awful physical feelings I have ever experienced I woke up in cold sweats for weeks afterwards, still believing I was in that Interrogation – but it also sits among the greatest experiences of my entire life. It surpassed my expectations as a show, and I discovered I had surpassed my expectations as a person. I had got what I needed out of it, and can look back today and feel both empowered as a person, and extremely grateful to the show's producers for giving me the opportunity

In the immediate aftermath, though, I found that I was left with this deeply empty feeling inside. I was pining for the next big challenge, but I didn't know where it could possibly come from, where to look for it or, really, if there was anything even out there that could ever match that experience in Chile.

I was soon drifting into a potentially dangerous and destructive mindset that would ultimately take another dramatic change of scene to break.

Chapter Sixteen
Down Under

It is hard to describe how much of a mental shift it was going from the climax of the *SAS: Who Dares Wins* Interrogation stage in Chile to being back home in Cumbria, all within the space of just a few days. At first everything was so exciting. I was ecstatic to be on the farm again. Fraser stuck to me like glue, my family could not have been prouder, and I was on the sort of high that was so intense, you'd be forgiven for thinking it would just go on forever.

But it doesn't, and then you come crashing back down to earth with a bump. I'd gone from an 18-hour interrogation, off the back of a 48-hour exercise with negligible food and no sleep, to a long-haul flight that I still didn't sleep on. The bank account of my body and mind was deeply overdrawn and, after about a week, the adrenaline coursing through my body subsided and I discovered just how mentally and physically shot I really was.

I was tired, but even after good rest and food, I still couldn't seem to find my usual natural energy and spark. 'Restlessness' would probably be the best way to describe the feeling that crept up on me as the distance between Chile and Cumbria lengthened. Before the show, farming our patch of Cumbria had been my absolute life and my greatest source of excitement, but now it all felt quite boring and repetitive. Was the rest of my working life forever going to be a variation on what I'd already seen? I knew there were always new things to learn – advancements in the industry, new breeds, new technology, new ways of doing things – but none of that felt stimulating in the same way as it had when I was on my farming apprenticeship, that steepest of steep learning curves, and none of that came even close to the raw exhilaration of tottering across a rope bridge high up in the Chilean Andes.

I guess what I was really thinking was: I know I'm lucky, I know this is the right job for me, I know I love my life . . . but really: Is this it?

What I wasn't capable of understanding at the time was that *SAS: Who Dares Wins* was always going to be a defining event in my life. It was impossible to explain to people that hadn't been through it. It was a one-off: utterly unique, irreplaceable and completely unrepeatable. Pretty soon, though, every day after a morning's work, I found myself just sat with a cup of tea in hand, gazing listlessly out of the window at the sheep in my field, and desperately wanting more from life. I had only had a taste of what it might have been like in the military, but already I could appreciate why so many soldiers struggled with civilian life when they came back to normality after a life of service. The lowest ebb came when the show actually aired in January, three months after I had returned home. The good feelings I'd had immediately after Interrogation were gone. I now felt I should have passed, having got to within touching distance of the ultimate success, and was more determined than ever to prove myself in another arena.

'Hey, Red Shepherdess, do you want to come out and do some lambing with us?'

The request came through my Facebook. It was nothing

out of the ordinary given my line of work; people made me random job offers all the time. However, *this* job was actually over 10,000 miles away.

I stared at it for a little while. *Could this be the answer to my problem?* I thought.

Realistically, I couldn't really make it work. My sister was getting married, I had my contract jobs and Brookside responsibilities to crack on with, and there was no removing the fact the job was on the opposite side of the planet. But that was exactly what appealed about this offer so much. Australia, the land of adventure. I had always wanted to go, and once this little seed had been planted in my head, it inevitably grew into much more.

My year dragged and the Australian lambing season soon passed. The farm had replied to my polite turn-down to say that I was always welcome out later in the year when their tail-marking and tagging season started up (when they pull in all that season's new lambs, and officially mark and ear-tag them to the farm, as well as dose them with all the treatments and vitamins they need). The more I contemplated it, the more I came to realise it wasn't quite so outlandish as it first seemed. That part of their season started up in our early autumn, a time when things here started to naturally quieten down. So, as the cooler, calmer months swung round, I found myself typing up a simple message to say: 'I'm up for it now, if you guys still are?'

They got back right away. The work was paid and the accommodation sorted. They also underlined that they definitely needed more hands for the tail-marking. The location sounded epic, so, in a classic case of not wanting to give myself any time for doubt to creep in, I applied for a visa the very next morning, was accepted within 20 minutes (Australia and sheep work – you'd be waiting longer if you were a rock star) and then booked my flights for the following week.

I know it might sound a bit mad. Accepting requests from strangers on opposite sides of the planet would be a decent beginning for the next smash-hit Netflix murder saga, but the shearing world is really small and, doing your due diligence, even on people thousands of miles away, is actually pretty straightforward.

Put aside the aforementioned evils of social media and there are a lot of really useful tools just for getting to know people. I've got close friends who I have never even met but will speak to all the time, and trust, and so many fantastic opportunities have come my way directly through my social media platforms. I knew straight away that this was one of those opportunities as, firstly, a single click showed me that this man and I had lots of mutual friends. A bit more exploration showed he was originally from South Wales in the UK, near an area I had worked often (hence the crossover with my mates), and a couple more clicks showed me his farm, his beautiful family, his wife and kids. 'Richard Davies' was pretty quickly elevated from 'random person on the internet' to 'legitimate work acquaintance', who I was very excited to meet and have the chance to graft alongside.

This was exactly the sort of break I felt I needed at the time. It was a new challenge in an unfamiliar landscape, but on a job I knew well enough to be able to bring something to the table. I was heading Down Under for five weeks. My plan was to earn some money for three and then spend two having a bit of an adrenaline-fuelled outback adventure. I wanted to prove to myself that it was possible to find a place where I could both earn and sate this burning need for excitement. Instead of just being a bit mundane, my farming skills could actually be the tools that helped me access the world, and all its edges; but I did harbour just one secret fear. What if I loved Australia so much that I didn't want to come back? I'd have to get Fraser flown out and quarantined for a new life, obviously, and my parents would have to get cover for all my work at Brookside – but the more I thought about Australia, this wild, uncompromising place filled with creatures that could kill you with a single bite, and all its limitless horizons, the more traditional Cumbrian sheep farming just seemed a little dull in comparison.

If I was feeling that way already, goodness knows how I would feel once I got out there and started. Mum spotted it in me right away. I tried to downplay it, but she knew. The Jackson family were holding their breath again.

Day one. It was Baltic.

I had thought the late Australian spring would be plenty warm enough for shorts and short sleeves, especially given I had flown direct from a Cumbrian autumn, but I was absolutely freezing. It turns out that, although the continental landmass of Australia is 70 per cent outback desert and the driest inhabited continent on earth, there is also this thick ring of greenery around the coastline, and even tropical rainforests in the northeast. Where I was, in the south-eastern state of Victoria, we were a significant distance from the Australia of my imagination – and, unfortunately, the Australia I had prepared for with the clothing choices in my bag.

My perception of what Australia might be like was absolutely nothing like the reality of where I was working. I expected isolated cowboy-inhabited hamlets scattered through a desert of sweeping red sands, dustbowl scrub, and snakes in every direction. Heat on scorching heat, lithe, lean-looking sheep and endless uniform vistas under a giant bright sun. Branx-holme ticked the 'one-horse town' box – it was a small Victorian township with one pub and one shop that doubled up as the post office – but it was all so grassy and green, cool, with big fluffy clouds and big blue skies. 'Quite, erm, Cumbrian,' I noted. I had come to the opposite side of the planet in search of the exact reverse experience to the one I had just left back home, and instead found a culture and environment that actually wasn't that dissimilar. I laughed to myself and, pretty quickly, was ordering extra jackets and a thick woolly hat for work.

Welshman Richard Davies had been a travelling shearer in much the same way so many Australian shearers chose the nomad's life, and had earned his crust cutting fleeces across the whole wide world of sheep. But, while working in Australia, he had met his wife, fallen in love, settled and married.

Rich and Ali sit comfortably among the loveliest people I have ever worked for. Rich is tall and thick-set: a proper big-boned and burly farmer. You could still hear his Welsh accent in there, but it was suppressed under an obvious Aussie twang and that weird Australian intonation where they go up at the end of their sentences, and make every statement sound a bit like a question. Both he and Ali went out of their way to make me feel welcome. They invited me to stay at their home, an offer I gladly accepted, and I ended up living with them, really as an extended part of their family, for the entire time I was on their farm.

Their home was this beautiful spacious wooden bungalow with a large veranda outside. Every evening I would leave my boots on that veranda, and then spend every morning in a state of near-frozen terror checking that a massive spider hadn't crawled inside during the night. I can't describe to you adequately how terrified I am of spiders. It wasn't just my boots I checked (and double-checked): it was the toilet seat, it was my clothes, it was all the farm vehicles I jumped into. The fact was, if I had ever actually discovered a deadly spider in my boots, or anywhere else for that matter, I probably would've crapped myself so violently that death would've come my way anyway.

During my time at Rich and Ali's they actually called in a pest-control expert to fumigate the entire house for spiders as, apparently, it was the season for a certain species to venture into people's homes. You might think the appearance of pest-control would be comforting for me, but no, I simply assumed that all the household spiders that had hitherto hidden from my view, embedded in their millions within the bungalow's walls, would now be encouraged to flee from their safe spaces and into my room, my suitcase, my clothing or my bed. There was a dark mark on the wall in the room I slept in, and there was not one night when I didn't wake up and mistake it for some giant arachnid about to reach down and rip off my face.

Rich and Ali's kids were adorable. Their youngest was a two-year-old lad called Ernie, who was super-sweet and would often dress up as a cowboy. Then there were two little blonde dynamos called Indy, who was about four, and Miley, who was six. From the day I moved in their kids were with me the entire time I wasn't working. Ernie was still learning his words and would occasionally struggle with pronunciation. He addressed me as 'Laid', I think, instead of 'Lady', and was absolutely obsessed with foxes. I discovered that fact when he shouted: 'Look at that big fuck!' out into the darkness one night. He spent a lot of his spare time either looking for foxes (which were a real problem on Australian sheep farms), watching fox videos, or practising his fox calls, which were basically these little squeaking noises.

He'd do his squeak, pause, then look you square in the eye with total seriousness and say: 'That's a fuck.'

After two days, the children were comfortable enough around me to come bounding into my bed first thing every morning. That was lovely at first, but after a week they'd graduated to thinking nothing of joining me when I was trying to go to the toilet. Secretly, though, I was just thrilled I had found a home from home. I absolutely loved those little guys.

Surprisingly, given the rough and tumble of the Australian farming lifestyle, the family's pets included an absolute queen of a sausage dog called Florence, a Jack Russell called Jack, and this cat called Tig, who could hold her own with any one of the farm's sheepdogs and took absolutely no nonsense off anyone. In a strange sort of way, being back there was a bit like being back at Derek's. Rich loved all his dogs. Alongside Florence, he had 18 working kelpie sheepdogs – a descendant of the British collie with a soft coat, pricked ears and a very athletic build – and four kelpie puppies. Every evening after work we would retire to his yard together, train the puppies, drink beer and talk dogs in the glow of the sunset.

I slid quickly into a routine of Australian family life and very hard work. I was in my absolute element and the weeks flew by.

I hadn't taken this job purely for a change of scene and some excitement. I wrote before about how much lamb is imported from the UK from this part of the world, but Australia is also the largest exporter of livestock on the planet, anywhere, and I was very curious to learn more about how they were pulling off such a trade. How could it be more cost-effective for any government to do a deal with Australian farmers over their own? Australia had to be doing things that we weren't, and it couldn't all be down to the size of their land and the numbers of their sheep.

Mind you, it was hard to overlook the sheer scale of sheep farming down there. Branxholme might have had some similarities with Cumbria, but this was Cumbria on speed. The average Cumbrian sheep flock hovers around the 400 mark, but Rich's farm had 18,000 breeding sheep and 23,000 lambs alone. Even then, his wasn't considered to be a true mega-farm – there were others that were double, even treble the size of his. They still needed to find sound cost-effective techniques to manage all of those animals though. Simply employing hundreds of shepherds, or having giant teams of dogs, or spending vast amounts of equipment wouldn't justify the amount they could afford to charge for their product. I knew there had to be more to it.

From my first day on the farm I realised everything here was built for purpose. It was efficient, streamlined and effective in a way that was light years ahead of what we had back home. There was no messing around tying hurdles together with rag-tag bits of rope to create a makeshift pen, or debates about what, where and when things were going to get done, if at all. In Cumbria, I would often turn up on farms and waste time figuring out how things worked, or discover vital bits of kit were missing or broken. In Australia, everything felt brand new and had been built in such a user-friendly way that all you had to do was apply a tiny bit of logic and common sense and it was blindingly obvious how things would and should work. Because they were working on such a giant scale, they could afford to build one big shed for single tasks, like shearing, for example, and everything required to do that one job was always inside, right where you needed it.

In the giant yards, the fences, runs and funnels were built in a way that recognised the natural behaviours and movements of the sheep too, and then our work stations were placed at points that would most suit their likely habits. For example, whereas all our British yards' fences and pens are generally built on straight lines, the Australians had these big horseshoe-shaped bends in nearly all their fences and runs. It sounds so simple, but what it meant was that when you were herding the sheep towards the yard, they weren't always facing a big white wall of sheep at their destination. It is an instant panic-starter when travelling sheep realise they are heading to entrapment, but here the sheep would naturally and instinctively follow the fence line's curve, and not see they were heading into a pen until they were already there.

They had simple funnel systems too, all with smooth tapered lines, unlike some of our British systems that have really odd corners that can easily trap sheep. They also built much longer 'races' – the single-file walkways you push sheep into to perform mass tasks like deworming, medicating or tagging and marking. Once I'd seen it, it was very obvious why: the longer the race, the easier and quicker it was to get in there and get those jobs done. You'd think with the numbers of sheep it would be non-stop shouting and bedlam, but simple design negated the need for more dogs, bodies and unnecessary extra expense.

It wasn't just the farmers and labourers that knew exactly what they were doing at all times – the dogs did too. The kelpies worked in a similar way to a collie, but they had adapted their working behaviours to life in these enormous sheep pens. I spoke before about how good collies could still create space in a pen filled with sheep by using their physical presence to cut routes through the middle of tightly packed flocks, but in Australia, the kelpies would actually just leap onto the sheep and run right across their backs, creating space as they crossed the sea of sheep, jostling them forward from directly above. It was impressive to see; they were absolutely fearless animals, and I had to admit it was a lot quicker than any ground-level movements and pressures we were doing back home.

I hadn't been too sure of kelpies as a breed before going to Australia. I'd encountered them a couple of times in the UK and they struck me as a bit too forward, a bit too in the sheep's face, not 'aggressive' necessarily, but forward enough to make the sheep feel overanxious and, I felt, create a collective nervous energy that could soon lead to pandemonium, if not constantly managed. In this environment, though, under Rich's expert guidance, I could do nothing but praise these dogs. I had never seen a sheepdog with such high levels of endurance either. In spite of the size of the farm, and the flock, Rich would only really use four dogs in a day. They would cover enormous distances, herd sheep in their thousands, and had an absolutely relentless work ethic. To my eye they worked as well as a collie in the fields, but in those mega-pens they certainly had the wood over my dogs. There was no way in hell you'd get Fraser springing over sheep's backs to get them moving along.

I had arrived for the tail-marking season. What this meant in practice was that all the ewes and their lambs that had been born earlier that year needed to be gathered off the fields and brought, in their thousands, down into the yards. Once there, the ewes and lambs needed to be separated, and the new un-marked lambs would first have their tails docked. The Australians didn't fanny about with rubber rings and multi-day waits for the docking to happen, they used a piece of kit called an 'iron', which was basically a gas-heated searing knife which instantly docked the tail and cauterised the wound. The freshly docked lambs were then vaccinated, given a 'CLiK' spray for extra protection against fly-strike, tagged in the ear, and had their weight and sex recorded.

All of these jobs were completed by individual workers, responsible for the individual jobs, all stood in one long line. With lambs being constantly fed down like they were on a conveyor belt, I could not quite believe how simple, yet undeniably efficient, the whole process was. On our best day we got through 1,800 lambs in a day, a fraction of the 23,000 lambs we worked through in total, but by Cumbrian standards that would have represented the entire flock on a substantial farm just in itself, and would have taken us at least a week. Upscale that to include the full 23,000 lambs on Rich's farm and, if we were in Cumbria, handling those numbers with our methods, we would probably have been flat-out tail-marking right through until the following lambing season.

I have to be honest: I had anticipated that farming on that scale would have seen a reduction in welfare standards. That caring for an individual sheep, to the same level that I will at Brookside, was impossible when you had one hundred times the stock. I had it in my mind, before Australia, that giving the best care always meant being hands-on with every individual sheep. For tail-marking in Cumbria, we will more often than not grab an individual sheep in a pen, having first isolated it with the dogs. Then one person will do all the docking, the vaccination and the measurements, before releasing it again. In Australia, having such a streamlined, almost factory-like assembly-line methodology meant the time the sheep were actually being subject to the stress of the job was far reduced. It was more impersonal, for sure, but that didn't always have to be a bad thing for the sheep.

Probably the biggest reason for the overall health of Rich's flock was that the sheep breed out there was simply a lot tougher than ours. The climate might not have been as varied as we have in Cumbria, but it could be much more extreme in terms of drought and dryness in the summer, and then the winter rains, which might bring widespread floods. The fact the Australians hadn't historically brought their sheep into sheds during the worst of their weather extremes, as we frequently do in Britain, meant that a lot of the weakness had simply been bred out of their sheep already. I still could never contemplate letting one of my sheep just die outside in the elements, but I did realise in Australia that there was an elemental truth to their farming. They handled their sheep a lot less frequently, and the sheep had a terrific amount more space and freedom, but with that, the sheep had to learn to fend for themselves a lot more. It was a closer application of the natural law of 'survival of the fittest', where the weaker animals simply do not make it to breed and pass on their weaker genes.

It felt a bit brutal and primal, but in modern times there was much more at play too. Australian farmers are guided a lot more by science and scientific profiling of their flock than we are. In Britain, I was still buying and breeding mostly from instinct and learned experience; a method nowhere near as accurate as the Australian application of actual empirical data that can inform you about the genetic history and hardiness of your sheep stretching back generations, at a glance. In Australia, the recording of their sheep's lives was just on another level to ours. Every aspect of a sheep's birth, life, growth rate and hardiness was recorded into electronic data-gathering systems, which allowed farmers to make purchasing and breeding decisions based on hard science, and not gut instinct.

It was undeniably a better way of farming a healthier flock, and today, I aim to be guided much more by science. Instead of simply buying sheep based on the 'look', I try and use data, gathered at source by technology companies like Shearwell, which can give me a complete profile of the sheep going back through their genetic line, and their history as individuals on the farm. We are very much behind Australia, but the more we record, the more accurate our projections will get.

The technology in Britain is better established with breeding tups. Already, for some individuals, companies such as Innovis have average performance figures for the growth rates, muscle depth, lambing ease, lamb survival and lamb vigour (how quickly it gets up) of all the children a sheep has ever produced, which can then be compared as a score against other tups that might be for sale that day. When the success of your whole farm hinges on breeding quality lambs, and spending thousands of pounds on the right sheep at auctions, it seems crazy that we haven't universally embraced more of the Australian-style science and data-collection facilities that are already out there. I still could never let a sheep die that I think could be saved, but in Australia, I learned to think twice before reintroducing it back into my breeding flock, and have already started making much more accurate records of all my animals with a digital recorder from Shearwell.

I don't feel my sheep are becoming just a number on a sheet, but I have to acknowledge that my farm is a business, and in these uncertain times I believe we should all be taking on board the things that can help make our businesses more secure. It is absolutely in my interest to make my flock as productive as possible, and I don't think embracing some technology is cheating on the traditional lifestyle of the shepherd, it's future-proofing it.

*

I had daily encounters with Australia's iconic wildlife almost the entire time I was away. Kangaroos were everywhere, and I discovered that if koalas were found when land was being cleared for development, their groves and habitat must be left intact. Often, then, I'd come across these random little patches of trees, bamboo or eucalyptus, with a beautiful furry little koala resident somewhere in the middle. I did (unfortunately) come across a few of the venomous redback spiders too – the beady little black bastards with the red splodge warning on their abdomens – and one of Australia's many deadly species of snake, which Rich quickly dispatched 'the Australian Way' (a spade to the head) so it didn't get in among his sheep. I also spotted a flock of emus, collectively called a 'mob', which seemed fitting given this bird can stand an intimidating six feet in height from head to toe, but by far and away my most amazing wildlife encounter came halfway through my stay.

We'd been at this big all-day dog trial with Rich showing some of his finest kelpies. He'd won a prize, so I remember we were all really buzzing as we made to leave. We were in two cars: myself, Ali, Ernie and Indy in one, and then Rich and Miley in the other. Rich had left before us, so it was pretty surprising when we pulled into the bungalow to discover he wasn't there yet. Not too much longer he arrived in the yard with Miley, looking a bit red-faced and flustered. 'Well, guys,' he began, 'we had an accident. Don't worry, we are okay.'

He explained he'd hit a wallaby on the way home, an occurrence that happens with amazing frequency during night-time driving in Australia; but unlike all the foxes, rabbits and hedgehogs that get flattened on UK roads each year, hitting a wallaby, or a seven-foot tall kangaroo, was a considerably bigger deal for the animal, your car, and the safety of all the passengers.

'Hannah, I've got something to show you,' he said.

I walked over, looked into the passenger seat, and there, cuddled up in Rich's jacket, was a tiny joey. It couldn't have been more than a couple of weeks old, hardly bigger than a bag of sugar.

'Oh my goodness, Rich!' I scooped it up right away and we all ran inside.

The baby wallaby had this super-soft, light-grey fur, quivering teaspoon-sized ears, and big black eyes. It had been in the pouch of the wallaby Rich had hit – and sadly killed – so he felt he had little choice but to pick it up and try to save its life at home.

Well, obviously I fell deeply in love instantly, as Rich always knew I would. We got a pillowcase for him to live in and I cuddled the joey all night long to keep it warm. The next morning we tried to feed it a little milk, and Rich called the local animal rescue charity. It turned out it was quite a rare species of wallaby, and the woman in charge of the charity requested its immediate adoption. She told me that had I not swaddled it that night it would definitely have died, due to its lack of fur and very tender age, but I can't say everyone wasn't a little gutted we were going to have to say goodbye so soon.

The kids were all in tears as we pulled into the yard of the animal rescue place. I felt I had to put a brave face on, and actually be an adult, as I explained to them all that this really was the best for our joey. I even forced myself to hand it over with a big smile, while secretly harbouring dark – and very silly – thoughts about how my dream to have a fully tame pet wallaby had just been thoroughly crushed.

The lady had a whole menagerie of damaged and recovering wildlife in her house: marsupials, other rare species of wallaby and kangaroo, and even a sub-species of sugar glider – a pint-sized and very cute species of possum that has a furry membrane attached to its hind and fore legs, allowing it to glide through the sky like a kite.

Our little joey made a full recovery with the proper care and, along with so much of the wildlife this charity looked after, its day soon came to be released back into the wild.

The sheep work was absolutely relentless. Breaks were swift – cram in a sandwich and carry on – but for the most part I was so intensely focused on the job I didn't even notice I was hungry. It certainly ticked any box I'd had going in that I was looking for a new 'physical challenge'. My job was usually to work the 'shedder', effectively a swing gate that would separate the lambs from their ewes as they funnelled from the pens and into the races. On top of that, I was checking the 'bags', i.e. the udders, of the ewes as they passed, to make sure there was no sign of infection or mastitis that might necessitate a dose of antibiotics.

Before I'd arrived, I had been expecting to work with a team made up of a hundred or so Geoffs, my sweary Aussie colleague back at Mike and Mary's place, but down on Rich's farm I felt he had embraced both the best of British farming and the best of Australian farming. It was reflected in the workers he'd employed. There was a lot more banter to be endured, and then served back hard, but it was all good-natured and great fun.

One of the central laws of any Australian sheep farm is that any major mistake must result in the purchase of an entire flat-pack case of beer called a 'slab'. On day one, it became clear to the Aussie lads lower down the line from my shedder that I was no greenhorn weak woman, and that if they were going to get any beer out of me it was going to have to be through under-hand tactics. At first it started with the very basic shouts of 'Oh, fucking hell, boys! She's fucked it again!' to try and goad me into looking up, telling them to 'fuck off', and then making an actual mistake through their distraction. It worked the first time, so that was one slab down and back-slaps all round for the lads, but I was never going to fall for that trick twice.

As the days, and then weeks, passed without error, they went properly below the belt. I have no idea where they acquired a smoke flare, but I knew about it when it got lobbed at my feet and started filling my whole work area up with a bright orange and impenetrable smoke. Absolute pandemonium descended, the shedding gate got left wide open as I ran from my station, and sheep and lambs started running down the line in their dozens. The lads were beside themselves laughing – to be fair, so was I – then we all realised just how many sheep and lambs had run through into the race and everyone had to down tools to sort out the massive mess.

So, I bought two slabs during my time. It could've been a lot worse.

The plan before I had left was to spend three weeks on the farm and the rest of the time travelling. In the end, I loved the farm and Rich's family so much that I stayed there for the full five weeks, right up to the day I absolutely had to leave. Australia made me fall in love with farming all over again. My quarter-life crisis was over. Farming wasn't just a means to an end, or a handy skill I could use to travel, it really was my whole life.

I discovered that there really were things that Australia was doing better than us. It wasn't just that they were winning contracts over British farms because they could control their prices by the sheer amount of meat they could produce: some of their systems were quite clearly better. We as British sheep farmers have to accept we can't just keep all of our traditions 'because that's the way it has always been' and expect supermarkets and buyers to pay more for our product out of some sense of patriotic duty. Brexit could mark the end of EU subsidies too, and then what? Are we going to ask for even more for the lamb we've produced, or are we going to be forced to look at some of the ways we do things? I think some of the Australian methods could work on the farms that employ me.

I've tried to bring some to Brookside already and I don't feel like I'm putting nails in the coffin of centuries-old traditions that go hand in crook with British shepherding. We have to adapt or face the very real likelihood that a lot of our farms won't be making the close of this century. Australia gave me the kick in the bum I needed to really focus on how I wanted to shape Brookside going forward. I wasn't going to just do the day-to-day any more, I had a huge opportunity to develop the land, to try and expand, to find better, more progressive ways to do things. Australia underlined I was absolutely in the right place after all – it was my mindset that needed to change.

I did realise, though, that I didn't ever want to farm on a scale that was anything like Rich's. I missed picking out the characters within a flock of sheep. That sense of individualism that comes with really knowing the life story of animals like 'Half Face' and 'Pecker' down on my own farm. Really, I just missed being able to sit out in a field and enjoy the simple meditative pleasure a shepherd gets from watching over their sheep. There was never any time for a pause for reflection while on the job in Australia. It wasn't a pure numbers game, but it was relentless enough to completely lose sight of the wood for all the trees.

In spite of all our quirks, I also found that I deeply missed the traditionalism that sits at the heart of the Cumbrian shepherding culture. I laughed when one of the Aussie lads started talking about our 'stone fences', which I realised meant our dry-stone walls that border all our fields and farms. Sure, wood and chicken-wire was quicker to erect, and more cost-effective, but there is an intangible romance to a hand-built wall that is several hundred years old, and is leaning in on a 5,000-year-old tradition. There is also an intangible romance to the sometimes unintelligible traditions that sit at the heart of much of what we do. Yes, we do need to adapt and modernise some aspects of British sheep farming, but many things are more valuable than money. Being in Australia made me very proud to be a British farmer too.

One of the best things that came as a happy side symptom of my time in Australia was that I took a proper break from social media. I still posted bits, but I was nothing like as prolific as I had been. Given the experiences I was having, you'd think I would've been all over it, but Australia taught me to live in the moment a lot more, and really to appreciate everything I had. It gave me the space to realise for myself that having a happy life wasn't going to come if I always spent my time conquering one thing and moving right on to something else. *SAS: Who Dares Wins* had naturally shaken up the order of things in my head, and that was actually okay and totally normal; but, in spite of everything, I now knew my true love and happiness was already waiting for me back in Cumbria. Spending time with Rich and Ali, and all their family, underlined just how much my own family really meant to me, and when my five weeks was up, I was very ready to get back to them all.

Chapter Seventeen

Danny, the Champion
of the World

Coming back to Brookside from Australia was very different to my return after *SAS: Who Dares Wins*.

I'd caught the train home from the airport and was closing in on my local Cumbrian station after a good 48 hours of travel. Outside the window the first fireworks were exploding into the night sky. It was Bonfire Night, 5 November 2019, and, I had to admit, I was feeling a little lonely.

I had really hoped to see Fraser when I arrived on the platform, but Mum had texted earlier to say they couldn't get him there, and then, because of her work, she couldn't be there either. My sister Holly had offered to come and pick me up instead, but just as I'd got into the final hour of the long journey from Rich's farm, she'd messaged to say that she too had been held up 'feeding the dogs'.

Feeding the dogs? Hellfire, Holl, I'd thought, staring angrily at my phone screen. *I've been away for five weeks and you can't even be here on time from down the road?* Obviously, I didn't say that, and instead did the classic British thing of texting back that it was 'absolutely fine', while my insides boiled.

It wasn't really Holly or Mum's fault that I was feeling a bit sorry for myself. Things had been a bit mixed up in my personal life for some time. It wasn't just deciding my true direction in life post-*SAS* that had made the Australia trip timely, it had also given me space to untangle all the knots in my love life.

Things between me and my boyfriend Mike had come to a head right before I'd left, and he was now my ex-boyfriend. We had been wanting different things from the relationship for some time, but I'd also come to realise that I'd developed serious feelings for one of my best friends, Danny. I knew Danny had felt the same way about me; many months before he'd said as much, but I'd badly knocked him back in an effort to stay loyal to Mike, and we didn't really speak that much afterwards.

Mike and I breaking up on the eve of my departure had meant Danny and I had been able to properly patch up our friendship. We had messaged each other and spoken a fair bit while I'd been in Australia, and now it finally felt like we were going to go somewhere other than 'just friends'. I thought we would definitely be seeing each other that night, at the very least.

I looked at Danny's message on my phone. He had written something about working late, and needing to pick up a Chinese takeaway. Had I definitely got it right? Were all these thoughts and feelings just in my own head? The train pulled into my station and I started gathering up all my things. Did anyone out there actually care that I was back?

I stumbled off my carriage, struggling with my bags, and took my first deep breaths of that Cumbrian winter. Then the whole world hit the pause button. There was Danny. Red-faced, with an armful of roses and eyes brimming with the happiest of tears. He strode over and scooped me right up, as I sank my own tear-filled face deep into the fibres of his thick jacket

*

Who the hell does that absolute gobshite think he is?

I had first laid eyes on Danny Gallagher 18 months previously, and I was less than impressed. Bizarrely, given he lived just down the road from Brookside, we actually met on Praia do Almargem beach in southern Portugal. I was competing in an international fitness competition that weekend, and had been stood nervously with my team, right before the start of the first event on the very first morning, when this Lancastrian loudmouth suddenly started bellowing over in our direction.

Surely he's not with us? I thought, but no, down the concrete steps and onto the beach jogs this absolute lump of a lad, striding out purposefully towards us all.

He was wearing these eye-wateringly shorter-than-short flowery shorts, a backwards baseball cap and singlet vest, and he also had these bulging muscles, a tattoo of a fox in a suit on his bicep, and an attitude to suggest he actually believed he was going to win this entire competition on his own.

It soon transpired he not only had no place in the contest, he was also clearly still quite drunk from the night before. Danny was just about the cockiest person I think I had ever laid eyes on, but he was, unfortunately, somehow still connected to our team, and had apparently arrived in Portugal to cheer us all on.

I had left the Lazonby Estate at the start of the previous winter and had needed to make some new friends that weren't immediately connected to my farming ex-boyfriend. In the new year I had started a relationship with a Royal Marine called Mike and, once lambing season was behind me, I joined a local gym advertising CrossFit classes.

CrossFit is essentially a high-intensity fitness programme that you do as part of a team, in pairs, or solo. It can be anything to stimulate almost any muscle in the body really, but it is always functional fitness, designed to help you with everyday movements, as opposed to just giving you a six-pack or a fashionably big arse. Exercises are really varied and can involve a whole mixture of climbing, lifting, running, swimming, throwing, hauling and carrying; but the thing CrossFit is most famous for is its 'box' – the name for the CrossFit gym you belong to – which is founded on the twin principles of being universally supportive, no matter what your level of fitness is, and completely non-judgemental.

That was why I joined. I've always loved keeping fit, but more than anything, I just wanted a social life and friends that sat apart from my work. Farming can force you into a very introverted space if you let it, and pretty soon you can find it has consumed every single aspect of your life. I wanted to meet a group of people that wouldn't always lead off a conversation about my favourite type of supplementary winter sheep feed. Also, as much as things were still going well with Mike at that stage, I only saw him every other weekend and he was frequently called away for military exercises abroad. Basically, I needed the company.

After just two weeks of one-on-one CrossFit training sessions, I had been asked if I would like to join the 'box' in a competition called 'Tribal Clash' based over in Portugal.

'Are you serious?' I replied, a bit taken aback. 'I've only been here two weeks and I don't know anyone from the box at all.' Apparently, they'd had a few drop out, and the coach explained I was definitely performing well enough to hold my own. I thought about it for a couple of minutes; I mean, it was just one weekend, after all, and at the very worst I'd definitely make some new friends. 'Okay,' I replied, 'I'll do it.'

It turned out that Danny was from our local CrossFit box

back in Cumbria, but I hadn't met him yet as I was still doing the beginner's one-on-one classes.

'Fucking hell, Danny,' piped up the team manager, 'where did you grow those quads?' Danny beamed down proudly at his own muscular thighs like a dad watching his kids in a primary school Nativity play.

I resolved to do my best to ignore him, his quads *and* his apparently giant ego. I needed to focus on what I had to do in the first event, which was now just moments away.

Danny was born in Bolton but his 'Gallagher' surname is a callback to his Irish roots. His dad had actually grown up in the rough-and-tumble Bronx neighbourhood in New York, as a direct descendant of the first generation of Irish immigrants who moved to work in the New World. Aged 16, his dad returned to the United Kingdom and (just as my great-grandad Bill had also done), he later found work in the booming construction industry. He did very well and by the time Danny was born, his mother and father owned and managed a string of 'Gallagher's' pubs across the Bolton area.

They were country people at heart, though, and Danny grew up very much a countryman. At the age of 11, instead of receiving the Motocross motorbike he'd begged his parents for, he was given a horse called Bobby. The bond was instant. Danny has always just instinctively understood horses. He says he got a lot of stick from the boys at school for being the lad who was into horses, but he quickly came to realise that, in the eyes of all the girls, being able to ride a horse was infinitely more attractive than anyone hoofing a ball around on some windswept football pitch. Danny's enthusiasm for horses significantly deepened after that particular epiphany!

His pathway to becoming a farrier – a professional crafts-man responsible for the shoeing of horses, their hoof care, and their balance – was not instant, but once he had taken his initial 12-week course and found someone willing to take him on for the full four-year apprenticeship, his life's path was set. His apprenticeship took him under the wing of a farrier based in Cumbria and Danny soon fell in love with the area, as much as he did his work.

Today, his job takes him from high-end racing yards, shoe-ing teams of horses worth hundreds of thousands of pounds, right through to individuals with a single horse in a single field that they've poured every penny they've ever earned right into. If Danny is anything, he is an everyman, capable of making friends with people in all walks of life. He's a self-made man, too, and as I got to know him better it became clear that both of our routes to the countryside had come from similarly unlikely urban backgrounds. We were both black sheep buck-ing the trends of the status quo, and it was definitely something that drew us together.

Danny is the kind of character who lights up a room. He isn't actually anything like as full of himself or 'cocky' as he can sometimes come across. Okay, maybe he is a little too cocky occasionally, but he is a pub baby after all, so very used to being in big social environments (and, to be frank, a lot of people from that part of Greater Manchester are hardly wallflowers). Being forthcoming is very much part of the culture, and it's not as though people from my part of the Wirral are shrinking vio-lets either. Danny just loves to make people laugh, he doesn't take himself seriously, and is happy to make himself a bit of a figure of fun for the far greater purpose of helping everyone get along and have a good time. But underneath all the bravado and brouhaha lies a man who is fiercely loyal to his friends, very thoughtful, and deeply sensitive to the feelings of others.

His sensitivity could not have been displayed more plainly than on the second, and final, day of that contest. Things had been going incredibly well. Our box had actually managed to progress to the finals for the first time in their history, and I had been performing well enough, despite probably being the most novice CrossFit athlete in the whole competition. The finals began with this massive open-water sea swim, though, and that was when I lost my nerve completely.

I couldn't quite believe the way I was feeling. By this point I had already competed in all of these exercises and challenges that I had absolutely zero experience with, but *swimming*? I had swum my entire life. In fact, when I had still believed I was going to work with orcas, I spent nearly *every day* in the open water perfecting my strokes, but there I was: absolutely terrified and rooted to the spot.

I think it had all gotten a bit too much. I suddenly felt exposed. I was in the final, surrounded by very fit people and the very fittest members of the gym I had only just joined. Pile on top the fact that this was a mass event with 120 people hurtling over each other to get to just one buoy, right out to sea, and it all felt like a recipe for my total humiliation. My hands had started to shake uncontrollably.

My boyfriend Mike was on the beach, but I got the standard military response from him: toughen up, block it out, perform. I had edged towards the start line on the sand and was fiddling with my goggles, looking around at all these much stronger, more competitive and highly confident women. I tried to calm myself down, but in truth, I was seriously beginning to consider pulling out altogether.

Just then I felt an arm around my shoulder. It wasn't Mike, it was Danny. 'Hey, Hannah,' he said, his eyes filled with sympathy. 'Are you okay?'

I had only spoken to him for the very first time the day before, and given how I'd felt when I'd initially seen him, you'd think he would be the last person I'd be spilling out my worries to. But he was also the *only* person right now, and the starting gun was mere minutes away.

I told him straight how terrified I was. 'Hannah, you'll do brilliantly,' he responded calmly, before telling a very kind white lie: 'Mike has just been saying to everyone what a great swimmer you are.' He gripped my shoulders with both of his hands. 'Forget everything else. All you need to do is swim out to that buoy and back, and if no one was here, you could do that with your eyes closed. You've absolutely got this.'

At the competition after-party, Danny and I found ourselves sat together on a log on the sand. We hadn't won the contest, but I'd come out of the open-water swim among the first in our box. I thanked him for being there for me, and we soon started nattering away like a pair of old friends. It was one of those conversations where you know you've really just clicked with someone. We talked about the competition, about our home lives and our work. He told me about how he had broken up with his girlfriend a few days earlier, and had come out to Portugal on the spur of the moment because he needed a complete break, a bit of fun, and some time away with his friends. Before we knew it, we had been away from the rest of the group for nearly an hour.

'Blimey, Danny!' I exclaimed, on realisation. 'I need to go and find my boyfriend!' I laughed, and we disappeared back to the party.

It felt then like I'd somehow known Danny my entire life, and not just for one weekend. From that moment forward we spoke or messaged each other nearly every day. We discovered we had an awful lot in common: we both clearly loved animals and the outdoors, we both loved talking, and we both always tried to extract the maximum amount of fun out of everything we ever did. We both had jobs where we worked for ourselves too, so there were often times when we could meet up for a coffee in the middle of the day, between him shoeing horses or me running some farm errand, and, pretty quickly that summer, our 'friendship' developed into a 'best friendship'.

Meanwhile, niggling issues with my relationship with Mike had started to grow into much more significant problems. I could have coped with the physical distance when he had to go away for work, but there was soon an emotional barrier too. That was a far greater problem.

Mike was trained to properly disconnect and separate himself from his home when he was at work. The military argue that it is a useful wartime skill. Soldiers that are missing their family aren't as focused on the job as they probably should be. A lapse in concentration could lead to a mistake, which could lead to a loss of life. Soldiers should compartmentalise their personal and professional lives into two clearly different boxes, I understood that, but what I didn't understand, until our relationship had properly progressed, was how *that* lifestyle and *that* mentality was maintained even when they weren't at war.

Things really came to a head in the build-up to *SAS: Who Dares Wins*. I felt the message Mike really wanted to impart was a reminder that it wasn't real war, in fact, it wasn't even real SAS selection. I got the impression that he thought it was all a bit of an embarrassment. He certainly seemed to shy away from telling some of his colleagues in the military.

You may be wondering why I stayed with Mike at all, and quite how our relationship could have lasted for almost a year and a half, but he was a lovely person at heart. When he wasn't away, he was different. He was loving, selfless and often a lot of fun to be around; but there were still moments, like on the beach that day, where he couldn't seem to access the empathy I needed, and I hadn't been prepared for the realities of what my life would be like when he was absent.

I now know that people in the military are conditioned to distance themselves; it is part of the job and, people who have grown up in a military family, know how to behave in a military relationship. Being in the military isn't just a job, it's an entire way of life. I loved Mike, and thought things might change. I wanted to work at our relationship and find a compromise but, ultimately, it just took me longer than it probably should have to realise that was never actually going to happen.

By the close of that summer, with Mike only being there intermittently, I developed feelings for Danny as more than just close friends. Instead of us seeing each other through pure convenience or coincidence, I started constructing ways of guaranteeing we met up. As Mike continued to detach, I was finding myself turning ever more to Danny for emotional support. Soon, I wasn't just looking to him if I was having a crap day, he became my primary source of fun, of laughs, of conversations about what was going on in the world, or on television, or just the latest dramas down on my farm or back in his stables.

We were sharing everything and I soon got to know Danny

inside out, but I also had to admit that I was starting to properly fancy the pants off him too, which, let's face it, is *not* normal behaviour in any friendship, and certainly was *not* what I wanted to be thinking about while I was still trying to make it work out with Mike.

I was absolutely not going to verbalise any of this. However, Danny – and I can't underline this enough – is an *incredibly* cheeky and flirtatious man. I knew he was feeling the same as me because, for reasons only known to himself, he had started telling my sister Holly that he was 'going to marry me one day' every time we'd got drunk together in the pub (much to Holly's annoyance).

That summer's end, we were paddleboarding with the Cross-Fit box, when Danny and I drifted off alone together. I deliberately pushed him off his paddleboard and, as he was rolling himself back onto the board and trying to stand up, he suddenly became very serious.

'Hannah,' he began, with his straw-blond wet hair stuck comedically to his face, 'I can't do this any more.'

My heart sank. I knew exactly what was coming, but I still tried to play dumb: 'Do what, Danny?' I said, with faux ignorance.

'This.' He motioned, open-palmed, with both hands. 'Carrying on pretending like we don't have feelings for each other, when we both clearly do.' He stared into my eyes. 'I need it to be more than "just friends", Hannah.'

I knew exactly what he meant. Of course I did. He didn't want us to keep treading water, pretending none of the feelings we had for each other weren't really there at all. *Shit*, I thought. I didn't know what to do. I had been in a state of pure perfect denial. I actually felt I *could* go on pretending these feelings didn't exist, forever. I don't know what I said. I think I just pushed him back in again. But the cat was out of the bag now, and it could not just go crawling back in.

Inevitably the subject came up again. I couldn't blame Danny, he was right, after all, but I was still steadfastly loyal to Mike and our relationship, regardless of any secret feelings I actually had towards my best friend. I explained to Danny that if things weren't to work out with Mike then it had to be because of problems within the relationship, and not just because of another man. I didn't think it was fair to Mike, and I didn't think it was fair to the commitment I'd made to fix our problems; but, in hindsight, it probably wasn't that fair to Danny either.

Danny was completely heartbroken. He'd been right about what he thought he'd seen in us, and he'd been right to broach it head-on, but me knocking him back so strongly had been unexpected for him. In his mind, that was us done. We couldn't go back to being friends after all that, and he withdrew from my life completely.

The whole time I was in Chile on *SAS: Who Dares Wins* I caught myself thinking about how much I wished I could tell Danny all about it, and not Mike. I wondered if Danny had moved on with someone else, or whether he'd *really* given up on us ever being friends again. I started to think about the reasons why I hadn't already just ended it with Mike too. Our relationship had continued in much the same vein emotionally, and I realised I was mostly now just sticking it out due to convenience.

It looks daft written down, but how many relationships out there are maintained just because breaking up would be hard, or complicated, or throw you into an unknown and uncertain place? Mike *was* a good guy, there was no dramatic reason why it wasn't working out, and many women would feel really lucky to be with someone like him. We weren't fighting all the time, he wasn't nasty in any way, we were just going in completely different directions, and every time I'd tried to voice it and sort it out, we'd always eventually end up right back at square one.

Once I'd realised things weren't going to get better, I hung on for the wrong reasons, such as: my best friend was married to his best friend; my parents got on well with his, and I loved his family too; and we had my sister's wedding coming up, Mike was an usher, and I didn't want our problems to affect her perfect day in any way.

Danny continued to ignore me after I returned home from Chile. It felt like we had broken up despite never actually being together in the first place. It was upsetting, it actually hurt like hell, but it was ultimately the best thing for Danny and me, and it certainly was the best thing in terms of me drawing my own conclusions about my relationship with Mike.

I thought Danny was probably gone for good at one point, but by the time it came to me going to Australia to farm we had patched up our friendship enough to compete together in the next CrossFit competition in Portugal, and all too soon my sister's wedding was over and I knew I had run out of reasons to stay with Mike. When the time came it was still really upsetting for both of us – break-ups are rarely nice – but I knew we had definitely run our course and that there was nowhere else for us to go as a couple.

Australia came at the right time. It sorted out my head after *SAS: Who Dares Wins*, it got me right back on track with my farming life, and Danny and I were able to start communicating with each other with real openness and no sense of guilt. Very quickly, I came to realise that one of the primary reasons

I wouldn't be staying in Australia was because I couldn't bear the thought of us being apart any longer. It was why I had felt so sure he would be there for me when I returned home on the train, and why I felt *so* disappointed when he had then texted to say he wouldn't be.

The outpouring of emotion on the platform, when I saw him that night, was like something out of a Hollywood film and undoubtedly the most cringe-inducing scene in Cumbrian railway history, but I was as happy as I think I have ever felt in my life. Clearly, my whole family had been in on the surprise too.

'You didn't really think I wouldn't be here to pick you up, did you?' he whispered gently in my ear, as we held our embrace. I pulled my face out of his coat and kissed him properly. It had been a long time coming, but, in that moment, I knew I was home.

Chapter Eighteen
Growing Roots

Danny and I were inseparable from that moment forward. Sometimes you just know it's right, and right then, we just knew. After such a long build-up, we were both completely ready to just be together properly. For the first few weeks we kept our relationship pretty quiet, though. After all the turbulence, it was just nice to be completely on our own, without any interference or opinions from anyone else outside of our immediate circle, and I wanted to be respectful to Mike too.

As time passed it became clear to me that Danny really was 'The One'. We each have our moments – we can both be stubborn people, and will occasionally wind each other up – but things have always been very easy with him. Everything from moving in together, to working together, to conjoining our families, to falling in love, has never felt anything less than completely comfortable and right.

Looking back, even though Danny and I could have got together a lot sooner than we did, and though I should've ended my relationship with Mike a long time before I actually did, I still have absolutely no regrets about how things worked out for us in the end. I believe the planets have a special way of aligning, and that big things only happen when, or if, they are truly meant to. You just have to keep yourself open to the opportunity, and then make sure you grab it with both hands when it really is time.

Danny never really stopped supporting me after that big open-water swim in Portugal. We are both each other's biggest cheerleaders, and always try to bring out the very best in each other. For the first time, I felt like I finally had a partner in life who really wanted to share all of his world with mine, and vice versa.

Danny and I met at a point where we were already comfortable in our skins and happy with our directions in life. There has never been any sense of competition, one-upmanship or jealousy. Danny is both my safe place and my biggest adventure in life. He is just as ambitious as me, and lends me all the extra courage I need to really push myself.

As 2020 began, we both knew we wanted to get going with building some proper foundations for our lives together. For me, that meant taking all of that momentum forward from my time in Australia and expanding Brookside. It wouldn't be easy. I first had to find that rarest of Cumbrian commodities – good farmland – and then it turned out 2020 had revolutionary ideas of its own.

In January, a highly contagious respiratory and vascular disease, called coronavirus, or Covid-19, began to spread out from central China. By the time winter turned into spring, coronavirus had expanded its range to affect virtually every populated corner of the planet. At the time of writing, ten months later, it has claimed over one million lives worldwide and caused most national governments to implement strict lockdowns: encouraging people to stay indoors, to wear facemasks, to remain socially distanced from each other, and to work from home wherever possible. The final scale of the damage it will wreak economically is still far from known, but it is already looking devastating for many industries, and has placed fresh pressures on farms and farming.

Our 'Natural Leaders' courses had to stop, and my contract shepherding ground to a halt after lambing time. Springtime school closures, to control the spread of the virus in Britain, meant farmers with families had to juggle full-time childcare with their work. Many farmers went into a holding pattern of doing their own basic animal husbandry and pushed back the bigger jobs until later in the year. Of course, no matter what happens to a population – warfare, environmental damage or a pandemic – people will always need to eat. Farms and their farmers ultimately had to find ways to keep working through coronavirus.

One of the few positives to have come out of the pandemic was that it sparked a new appreciation for the nation's food producers. At the start of the UK-wide lockdown in March, supermarket shelves emptied of essentials for the first time in many decades. People were panic-buying, and our supply chains were exposed for being over-reliant on overseas producers. Being forced to stay local also forced people to buy and think local, and I believe that in turn helped consumers become more aware of the work of the food producers on their own doorsteps. In our area it has been reflected in a sales spike among farm shops and smaller independent food retailers. I too was able to expand my own boxed-lamb business outside of Cumbria, making deliveries nationwide for the very first time, and plugging some of the holes left in my income.

The year 2020 will forever be globally remembered as a time of great sadness and disaster. My sister Holly is a hospital doctor and worked the whole way through the initial wave of coronavirus infections that first spring. She quarantined herself from our family and her husband so she could do her vital work without risking the health of any of us – just as so many key workers in our National Health Service did – and was then on the frontline as people passed away from this horrendous disease, separated completely from their families.

Many solid businesses had to fold completely, thousands of people lost their jobs, and the long-term projections now suggest we are looking at a long period of austerity to pay back all the economic debts incurred when the nation was forced to largely stop working.

We have stayed well at Brookside so far, and the quieter year has given me the time I needed to really focus on our farm without getting distracted by all my other commitments. I feel more fortunate than ever to have a relatively stable job, and grateful in ways I could never have imagined that I took the decision to move away from the city and live out in the countryside, but I also recognise that I have been afforded a privilege, and slice of luck, that many are sadly unable to access. This year, many more times than I can ever remember, I have found myself counting my blessings.

Before the extent of coronavirus in the United Kingdom had been fully realised, I was extremely fortunate to get hold of some more land for Brookside. I was having a conversation with my friend Mark – who, incidentally, was the field specialist I'd called on during the stubble-turnip disaster at Lazonby – and casually mentioned I was looking for more acreage, asking if he could let me know if he came across any fields for rent or sale while on his travels.

'Well,' he replied nonchalantly, 'I've actually got 32 acres for rent myself.'

My jaw dropped. 'I'll take it,' I fired back immediately, agreeing on a price before I'd actually even seen it.

It turned out to be a great decision. The location and the grass were both perfect, and it more than doubled the amount of land I could farm. Soon, I was buying new ewes and lambs from old friends, including Derek Scrimgeour's daughter Rachel (who is now happily running her own farm), and effectively trebling the size of our flock.

In 2019 I lambed 60 ewes at Brookside; with the expansion in 2020, I now plan to lamb 200 ewes in the 2021 season. Mark has also cut us a deal on a field of stubble turnips – so we are going to get all of the turnip goodness and none of the drama of actually growing the bloody things, which will be hugely beneficial to the nourishment of my new flock this winter.

Having got the ball rolling with the new land at Brookside, a lot of what I wanted to achieve in the short term actually happened at a real pace. As much as I have always loved the varied nature of all the animals I've kept on the farm over the past few years, I had to accept the focus must now be on our lambs, making food, and earning money. I moved my beautiful Highland cows to a local family farm to free up more grass for my much larger flock, and, instead of having a slightly mixed sheep flock that included Herdwicks and Hampshire Downs (that needed to be kept in different breeding groups), I have streamlined to focus on breeding just our most successful and prolific North England Mules. I am also using my handheld electronic recorder from Shearwell every day, scanning ear-tags on all the sheep I own and recording everything about that individual, from its medical records to its movement (from auction, to field, to abattoir), right in line with the Australian systems I'd seen at Rich's. I've just released my first Innovis tup to breed with this season's ewes too.

The only slight break from my otherwise laser focus on Brookside's efficiency in 2020 was Danny's slightly unusual habit of purchasing pairs of random crossbred black sheep and sneaking them into the flock. I guess all lads have their hobbies, and I'm sort of responsible as I did get him his first pair, but he definitely knows he is being *very* cheeky indeed.

That aside, as the year and farm has extended, I have realised just how much I love working with Danny. He has always managed to bring the sense of fun, adventure and excitement that you need in this job. No matter how much you love farming, the time to fix that post, build that pen or clean out that barn always comes around eventually – and it always completely sucks – but when you've got someone like Danny to go and do all the crappy jobs with, well, it's not half as bad.

Danny continues to work exceptionally hard in his full-time job as a farrier, but he also lambed his first sheep this year, he bought his first sheepdog, and now he is experiencing his first breeding season. As he learns, not only are we getting the work done a lot faster, but I am finding myself tapping right back into all those good feelings I once had when I was learning too. Danny really is reminding me all over again why I think sheep farming is still the best job in the world.

All that said, neither of us want to build the farm to a stage where it consumes our every waking hour. We want Brookside to be as successful as possible for all of our family, but it can't be all of our lives either.

There are too many things we love outside of farming: hill-walking, wild camping, climbing, paddleboarding, skiing or, to be frank, just having a beer and a chat when all the work is done in the tack room or sheep shed. If your farm gets too big, you lose your connection with your animals, your loved ones, and your sense of your own self. So many times, I've been sat in the pub with other farmers, and heard them speak regretfully about how they never saw their children grow up, just because of their workloads.

If we are blessed with children one day, I want them to get all the good things you can have as a kid growing up on a farm – appreciating nature, appreciating where your food comes from, and being surrounded by animals – but I also always want them to know that there is a whole other world out there. Preferably, we want to be able to show them it ourselves.

Danny and I are a true partnership. I've got Fraser but, until Danny, I always lacked that human companion to farm together with. I have to recognise we are quite unusual. He has no ego in being told what to do by me, but he is also a natural farmer, who is learning fast and will soon be on equal terms with me. When that happens, I will never revert to being a 'farmer's wife' to Danny, nor will he ever be just a 'farmer's husband' to me. We are both part of the same team; a fact never truer than when we ran back into my old foe.

I was walking through Brookside with Danny and an old farming friend. We were just chatting about the things I'd recently put in place and the changes we'd made, when my friend stopped in his tracks and remarked, 'I'll tell you what, since Danny's come along it just seems like the farm's really pushed forward.'

I couldn't quite believe what I was hearing. This was a very good friend who knew full well how hard I'd been working over the years, and exactly how much work I'd done alone just to get us to this point. Was he really saying that our success was purely down to Danny? I was very offended and about to give him a piece of my mind as he continued to list all the things that had happened since Danny's arrival, when Danny stopped him dead in his tracks.

'Well, it's nothing to do with me,' he flatly replied. 'Everything here is Hannah's work and vision. I just help her.'

My farmer friend didn't really know what to say. It was probably the first time he'd ever seen a man in agriculture defer all praise to a woman. I could not have been prouder of Danny, or us, that day. There really was nothing else to say.

Dove Crag mountain sits in the Eastern Fells of the Lake District, almost in the centre of the National Park itself. On the eve of my twenty-eighth birthday, Danny and I were sat together on its summit, watching one of the most glorious sunsets I think I have ever witnessed.

We were in absolutely no rush to leave. Our accommodation for the night was close and 'pre-booked', inasmuch as we had already laid out our bags and sleeping bags to claim floor space in the 'Priest's Hole', a small cave hidden just off the mountain's peak. We were huddled up in our own little world, chatting, laughing and toasting the ever-lengthening shadows that spread across the landscape beyond our boots, when Danny suddenly came over all serious.

I had thought about this moment a lot, pretty much ever since we had got together, but now it was actually happening I felt my world descend into a sort of sublime surrealism. Everything slowed down and felt like it was floating.

Is Danny really about to do what I think he is about to do?

He had stepped away from me and was already talking.

'I'll just begin by saying that this year has been absolutely amazing. The fact that we finally did get together after being best friends for so long. I know I might infuriate you and wind you up, but you are everything I want in life, and when I wake up in the morning, I smile when I see you there, and I smile again when I go to sleep, just knowing you will be there. It is the best feeling in the world, and . . .'

I could feel myself slipping into a different dimension. It was such a complete moment, I just wish I could have captured all of it – the emotion, the feeling, his words and that place – and popped it all in a jam jar to look at forever more. But he was down on one knee now and my overwhelming feeling was one of total shock. He'd just asked me to marry him, and now he was holding out a beautiful ring.

I was plunging my face into Danny's coat once more, like we were right back on the train platform where it had all begun. 'Is that a "yes" then, Hannah?' he asked sweetly, and I eventually managed to stammer out the correct three-letter word from beneath my emotional blanket. Gently, he slipped the ring on my finger. Danny, I already knew, beyond any doubt, was the missing piece I had been searching for my whole life.

I'd had absolutely no idea that he had planned to ask me that day, and I'd been so wrapped up in celebrating my birthday and climbing the hill that, when he did, it had taken me by total surprise. It turned out Danny's original plan had been to pop the question in the cave itself, but as we'd arrived that afternoon he'd discovered a lone camper already set up on the rocky floor. (I found out later he had texted my mum in a panic, and she'd replied: 'Pay him to leave or just push him off the mountain.') By the time we'd returned to the cave, though, with a ring on my finger and mascara streaming down both my cheeks, the solo camper had been joined by two more hikers and, with the pressure now off, the night passed by in a happy engagement party-blur of sloe gin, campfire songs and the very best kind of foods: those cooked with love from just one pan, and eaten with one shared fork.

We walked off the mountain the next day under the most bliss-filled blue skies and sunshine. I allowed myself to think that Nan had pulled some strings somewhere up there. I was only half-kidding – I mean, it nearly always rains on my birthday – but, an hour or two later, while in the car coming back to Brookside, I looked across the sky and there, right above our farm, was the most perfect halo of cloud. I had never seen anything like it. With a darkened central hole and a thick bright white ring, it was fully encircling our home.

I made Danny stop the car. That time, I absolutely *knew* it was her. I could feel it in the very fibres of my soul. It was her nod, her approval, her blessing. I told her I loved her, I missed her, and that I loved Danny Gallagher with every bit of my heart.

She would have absolutely loved him too.

Chapter Nineteen
Woman's Best Friend

I'm full circle, back on the gather, right back at this book's beginning. The sheep had emerged from the blind side of their fell and were successfully ensconced in their sorting pens. In spite of the weather, it had been a textbook gather.

There was a very different sound to our work now. The mood was upbeat, jocular, with piss-taking of the we-got-away-with-it-again variety. Swear words sang out over the backs of bleating sheep, the fraught part of the job left behind somewhere up on the whistling moors the other side of the hilltop's spine.

Five farms, and five farms' worth of farmers and dogs, piled their way through the full flock, trying to wrestle sheep into the pen that corresponded with the coloured mark dyed deep into their sheep's fleece. It was the mark of the farm to which that animal truly belonged. My farm's mark was absent here today, but at least I had one. I truly belonged now too.

The gathering season was drawing to a close on the most splintered year of contracting work in my memory. It wasn't just the pandemic's fault that I'd worked far fewer contracts: Brookside's rapid growth was always going to pinch my time from somewhere and, as much as I loved being a contract shepherd, I had to accept that was for the best.

If this year had taught me anything, it was that I couldn't

rely on contracting to always be there for me because, for as long as I did, I would be forever placing my faith in the strength of other people's businesses to sustain my own. I had to believe in Brookside now, just as I'd once had to believe I could make it as a shepherd and a sheep farmer.

Fraser turned ten this year. Senior in dog years, but ancient when weighed against the work ploughed deep into his paws and muscles. Eight years of hard graft at my heel. How many Mount Everests has that dog summitted in his lifetime? And how many more did he have left in him?

Having his protégée, Butch, up there, had definitely made a difference. I could sense he didn't want to be shown up on the gather. That he wanted to prove he still had it, that he remained the top dog and the pack leader – but we both knew his powers were finally and firmly on the wane.

Ever so slightly, his confidence had begun to diminish. Last season I watched him square up to a big tup in the pen, a move he had executed so many times before. A move that always ended with the sheep, no matter what their size, backing down, heading back and performing Fraser's bidding. But that tup, that day, had called his bluff.

He could smell the fear in Fraser. It was a throwback moment between those two mammals, a snapshot into a wild that we try to suppress under their modern-day domesticity, but it was still there, and in moments like that it often comes erupting out. Fraser blinked, took a half-step back, surrendered space and psyche to his adversary, and the tup immediately seized on the momentum. In a flash, the tables had turned completely. Soon, it was Fraser who was cowering, having just received a solid head-butt from that belligerent sheep.

I rushed over. He was okay physically, but psychologically a line had been crossed. From that point forward, Fraser stopped standing up to any sheep that displayed anything other than immediate submission. It didn't make a huge difference in his ability to work day-to-day, he still operated to all the other commands, but we both knew it was there. There was a chink in his armour, a weakness that could not be repaired.

Time and again I would watch him go into a pen and meet the resistance of other sheep, and time and again, he would submit, his head unbloodied, but bowed. Like a former world champion heavyweight boxer, stripped of their title, advanced in years, forced to take a knee when faced with far hungrier and fresher opponents. It is only the mercy of the fighter's corner-man that saves their legacy and extracts them from battle with the waved white towel. Likewise, I knew I needed to try and protect Fraser, but I still willed him to find a semblance of his old shape and sense of self. Neither of us was really ready to accept the truth.

It wasn't until I moved my new flock onto the new land this year that I knew we had to face reality. My sheep filled the road's width as we drove them together towards the gate of the field. As we approached the point we needed to turn them in through the gate onto the virgin grass, I'd sent Fraser up the side of the flock to block the road ahead. His position and presence should have easily halted the sheep in their stride, leaving them no option but to pass through the open gate, but after my call he'd turned back before he had even pro-gressed halfway through the flock. I sent him once again, he returned to me once more, and then he just sat down. He hadn't misunderstood the command, he wasn't being belligerent – he'd lost his nerve. He had thrown in the towel.

In the end I had to sprint up to the head of that column of sheep, physically stop them from progressing up the road, and force them into the field myself.

A new stage of Fraser's life had begun. Sheepdogs aren't sheep farmers, they are high-performance athletes. They burn bright and hard for a window of time and then they fade. That's not to say they can't ever find the whispers of previous form again, but you have to accept you can't absolutely trust them to take the solo lead under pressure, especially when you have young blood like Butch pushing them for their place.

I needed to shield Fraser now or risk ruining him as a dog. I didn't want to work him into the ground or place him in situations where we both knew he would be uncomfortable. What I feared, apart from him being seriously hurt, was that we could one day meet a point where I had eroded the most vital of connections between a sheepdog and their shepherd: the bond of our trust.

Heavy hands gripped hard onto wet wool, as dogs isolated and pressed rogue running sheep. The sorting of the flock among the five farms was almost complete. Up there on the wild open fell, those animals had experienced their final taste of freedom and foraged foods for the season; now it was time for their return to the civility of the lowland fields. There they could graze on what was left of the good grass, before stubble turnips and supplementary feeds would see them all through the winter and on to lambing time. No matter what happens in the wider world, the basic rhythm and cycles of the traditional shepherd's life are absolute and unchanging. In times of turbulence you come to learn there is a real comfort to be found just in that.

Hill sheep are often 'hefted' to a landscape. It means that

even when they have mile upon mile of space to roam into, sheep from a particular farm and fell will stick to precisely the same spots as their family did before them. It is an intensely romantic piece of learned behaviour passed first from shepherd to sheep, and then from ewe to lamb, over many successive generations. For those sheep, the grass isn't greener on the other side; a hefted flock will remain precisely where they have always been, and fences or walls are no longer necessary to keep them in their place.

The idea of being 'hefted' to the land is a concept often co-opted by traditional sheep farmers to describe their own close relationship with where they come from. Their life, land and culture is ascribed to them through successive generations of farmers from within their own family lines, often stretching back several hundred years, to an area perhaps not larger than a couple of farm buildings and a few miles of fields.

It was a feeling I could not quite access as a first-generation farmer, but this year, finally, I began to feel hefted to my own patch of Cumbria. I can run my fingers through Fraser's fur, along our stone walls, or through the wild grass in Brookside's meadows, and just know this is where I am meant to be.

But no person should ever feel that their life is preordained from birth. Being hefted as a person is not the same as being hefted as a sheep. The modern world should be viewed as a 'fenceless' landscape, where there are opportunities to change your direction and life's path if you feel the same burning need as I did the day I first saw Larry the lamb.

No matter what you're told, your background – whether it's your gender, your race, your religion or your class – should never be the thing that stops you from realising and then achieving your dreams. I had no birthright to be a sheep farmer.

I made this life for myself. I know I have had my share of luck, but I also know that I have worked hard and not compromised on the voice in my own head that repeatedly told me that I *could* do it, that I *could* farm, that I wasn't just some townie Scouse woman whose roots could never be moved and planted somewhere else entirely.

I feel my roots extending into this earth now, and sometimes I even feel that this earth itself is accepting me into its embrace with its own subtle serendipity. Danny and I received the ancient deeds for our new house this week. It sits opposite Croglin church, so its modern 'Church Cottage' name makes sense, but there, inscribed in the 300-year-old paperwork, lay its original name. It is the Red House. The Red House for the Red Shepherdess and Danny, right across from my sister Holly and her husband James, and my mum and dad, who have finally moved to Brookside full-time. I hope there may even be space for my sister Caitlin, if she ever wants to move to Croglin one day. We are all meant to be here.

In the future, I want to expand the flock and the farm just a little more, but I also want to open Brookside to the public. Sheep farming is not a closed book, and if my story is to mean anything, I know I need to keep doing my bit to bring people into our farming world. Alongside all the stories I post to social media, I would love to have farm stays at our place, with simple instructional sessions about the realities of sheep farming. It has never been a more important time to bring the public on board to back British farmers, but on a simpler level I just want people from towns and cities to come here, to experience this beautiful corner of the world, to breathe the fresh air, and to get the grit under their fingernails. One day, even, I would like to imagine that a young person, who may not have ever been on a farm before, might stand in our fields and witness something life-changing for themselves. That they too might find the inspiration they need to turn their back on convention and obligation, and discover their own pathway to farming.

Danny, Butch, Fraser and I led one of the five flocks off the mountain in one long line. I could see their new home – their farm's uniform fields, its yards and sheds – but, as we crossed the threshold between fell and field, I took a moment to pause.

Fraser dropped back by the heel of my boot as Danny and Butch pressed the sheep forward alone. Both already so self-assured and so capable, you would never have guessed it was their first time up on a gather.

I felt that things had changed in just the time it had taken to complete that day's job. I looked deep into Fraser's eyes and rubbed his cheeks. For so long it had been just me and him. Our lives intertwined, enmeshed, deeply woven, sometimes tangled even. I didn't want to think about Fraser being old and I didn't want to think about how the special balance of power between us might be slowly shifting in my favour. Instead, I stood still, I breathed deeply, and I looked ahead.

Fraser still has many good years left, but we are no longer alone. I could see Danny and Butch, I could see their work, and I could see all our futures clearer than ever. It was bright. It was Red.

Acknowledgements

I'm not sure how I'm going to get through this without crying or without it being longer than the book itself. This book and journey have been made possible because of so many amazing people that surround me and support me more than I could possibly explain.

Thank you, Mum, for being the best role model a daughter could ever have. For being my voice of reason, my shoulder to cry on, my advice giver. For believing in me more than I do myself and for showing me how to break through glass ceilings. Thank you for always saying yes to pretty much every animal I wanted (apart from when Danny brought the hound pup home). If I grow to be half the mum you are to my children, I'll be so happy.

Thank you, Dad, for always teaching me as your little girl to reach for the stars and dream big. For taking me to do all the 'boy sports' when other dads were taking their daughters to ballet. For always proudly telling the world about the journey of the Red Shepherdess (at every opportunity). That sense of pride has pushed me beyond what you could ever imagine.

To you both, thank you for showing me true love is a partnership; it's compromise and it's growing, learning and loving together as best friends, not just husband and wife. Thank you for your infinite support, encouragement and love, and for taking a big leap with me in embracing this farming life.

Danny, the moment you came into my life it was never the same again. I saw a piece of you inside me and I couldn't let go. You were my best friend first and now you're my whole world. Thank you for picking me up when I'm down and for being my biggest cheerleader. For loving me unconditionally, and for jumping headfirst into this journey with me. You brighten up every day and I'll spend this lifetime trying to show you what you mean to me. I love you, no buts.

Fraser, through this journey you've been my constant, my right-hand man and the backbone of Team Red. Thank you for always standing by my side, for teaching me the importance of loyalty, love and friendship. I wouldn't be the person I am today without you. Words can't describe what you've brought to my life. You're my one in a million, you're my boy.

Nan, thank you for guiding my dreams from such a tiny age and taking me to every petting zoo and farm. For showing me kindness is what makes the world go round although I've never found anyone with a heart as beautiful as yours. I miss you every day. Thank you for reassuring me you're always with me. I hope you're proud.

Team Red, I am forever in your debt. Thank you for making the good days the best and the hard days easier. Thank you for always bringing a smile to my face and making the journey so much easier with you all by my side. I don't deserve any of you, but I love you all.

Holly and Caitlin, thank you for always fiercely having my back, for being my constant cheerleaders and for always being brutally honest (even when I don't want to hear it). We've always been three best friends as well as sisters and I can't thank you enough for everything you've done. You're my rock.

Ethan and James, thank you for being the annoying brothers

Acknowledgements

I never had, but also fantastic friends, for everything you've done to help on the farm throughout these years and for supporting me when I've jumped ship to head around the UK/world to farm.

Bec – past, present and future, you've been there through it all. Thank you for being my best friend for the last 24 years, no matter the distance or what direction life has taken us in.

Derek Scrimgeour and family, thank you for giving me the opportunity that changed the course of my life. For your patience and kindness when I knew nothing or no one at all. For looking past the stereotypes that many couldn't. You taught me so much about farming and dogs that I'm forever grateful for.

Social media can be hostile and lonely at times so a special thank you to James Robinson, Will Evans, Peter and Paula Hynes and John Kelly for having my back and lifting my chin if my head went down. You'll never understand what your support has meant through these years.

Thank you to my industry, especially to the NFU, NSA and these individuals: Minette Batters, Stuart Roberts and Phil Hambling. You've given me opportunities, mentored me and answered my questions along the way. I promise to learn more about politics!

Thank you to those who've seen me as nothing but a farmer. To me that's one of the greatest forms of respect. Special thanks to the Bowman family, Brough family, Rebanks family, Pickthall family and to Mark Holliday, your support means the world; you've employed me, been there for me and rented me land. Without people like you, I wouldn't be where I am now.

Will Millard, I'm forever grateful to you for putting my voice onto paper so perfectly and helping me tell my story in such a powerful way. What we've created together is one of my proudest achievements and I can't believe I'm now holding this book in my hands. Not only have you become an amazing friend, but you've helped me revisit and heal some old wounds as we've gone through this book – you've brought me peace (thank you to Grace for the guest appearances at these times, bringing smiles all around). I wish I had a bigger word limit to express my gratitude. Thank you for everything, Will, from the bottom of my heart.

Thank you to Ebury for this opportunity to tell my story and share my life's journey in so much detail. I feel so lucky to be part of the Ebury and Penguin family. Thank you so much for your belief and guidance along the way. A special thanks goes to Michelle Warner and Sara Cywinski, my amazing editors, who've worked tirelessly behind the scenes (probably more than I realise) and for believing my story should be told and heard further – I'm so grateful for this.

Thank you, Jenny Heller, my amazing agent, for believing in me and supporting me with this journey. Thank you for always fighting my corner, and for just getting me from day one. And of course, thank you for sharing your love of cute and crazy animal videos with me daily!

Thank you to those who tried to knock me down: the truth is you made me stronger.

And finally, the biggest thank you of all to every single person who has followed this Red Shepherdess journey over the years. I feel so incredibly lucky and humbled by all your love and support. You're the reason this book has happened, you're the reason I push to be better. Thank you on behalf of myself and Team Red.

Keep it real every day, keep it red! x